P9-CSE-288

The ART of WAR

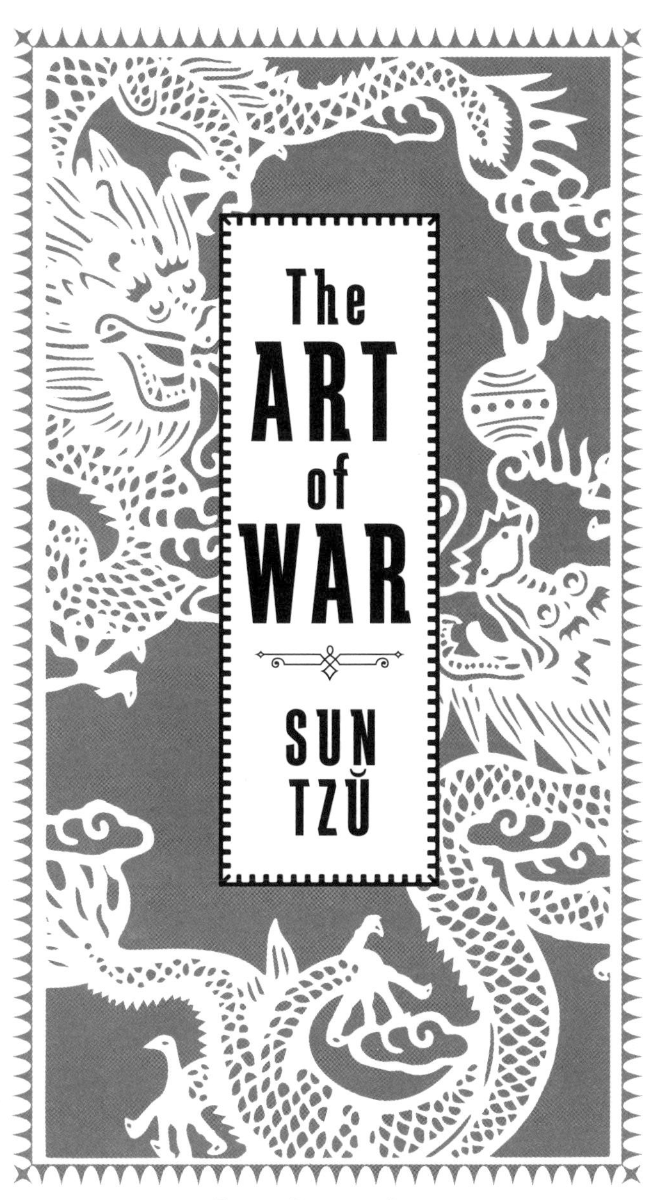

FALL RIVER PRESS
New York

FALL RIVER PRESS

New York

An Imprint of Sterling Publishing Co., Inc.
122 Fifth Avenue
New York, NY 10011

FALL RIVER PRESS and the distinctive Fall River Press logo
are registered trademarks of Barnes & Noble, Inc.

ISBN 978-1-4351-5869-6

Manufactured in China

14 16 18 20 19 17 15 13

sterlingpublishing.com

Cover design by Patrice Kaplan
Cover illustration © Hung Chung Chih / Shutterstock
Endpapers © my may day / Shutterstock

CONTENTS

Introduction 7

The Art of War

I. Laying Plans 47
II. Waging War 51
III. Attack by Stratagem 54
IV. Tactical Dispositions 58
V. Energy 61
VI. Weak Points and Strong 65
VII. Manœuvring 70
VIII. Variation of Tactics 75
IX. The Army on the March 78
X. Terrain 84
XI. The Nine Situations 89
XII. The Attack by Fire 99
XIII. The Use of Spies 102

Endnotes 106

A Note on the Text

The text for this volume has been adapted from Lionel Giles' translation of *The Art of War* (Luzac & Co., 1910). Endnotes have been adapted from Giles' annotations for his translation. Paraphrases are indicated by bracketed text. Elisions are indicated by ellipses.

INTRODUCTION

Sun Tzŭ Wu and his Book

Ssŭ-ma Ch'ien gives the following biography of Sun Tzŭ:—

Sun Tzŭ Wu was a native of the Ch'i State. His Art of War brought him to the notice of Ho Lu, King of Wu. Ho Lu said to him: I have carefully perused your 13 chapters. May I submit your theory of managing soldiers to a slight test?—Sun Tzŭ replied: You may.—Ho Lu asked: May the test be applied to women?—The answer was again in the affirmative, so arrangements were made to bring 180 ladies out of the Palace. Sun Tzŭ divided them into two companies, and placed one of the King's favorite concubines at the head of each. He then bade them all take spears in their hands, and addressed them thus: I presume you know the difference between front and back, right hand and left hand?—The girls replied: Yes.—Sun Tzŭ went on: When I say "Eyes front," you must look straight ahead. When I say "Left turn," you must face towards your left hand. When I say "Right turn," you must face towards your right hand. When I say "About turn," you must face right round towards your back.—Again the girls assented. The words of command having been thus explained, he set up the halberds and battle-axes in order to begin the drill. Then, to the sound of drums, he gave the order "Right turn." But the girls only burst out laughing. Sun Tzŭ said:

If words of command are not clear and distinct, if orders are not thoroughly understood, then the general is to blame.—So he started drilling them again, and this time gave the order "Left turn," whereupon the girls once more burst into fits of laughter. Sun Tzŭ said: "If words of command are not clear and distinct, if orders are not thoroughly understood, the general is to blame. But if his orders are clear, and the soldiers nevertheless disobey, then it is the fault of their officers.—So saying, he ordered the leaders of the two companies to be beheaded. Now the king of Wu was watching the scene from the top of a raised pavilion; and when he saw that his favorite concubines were about to be executed, he was greatly alarmed and hurriedly sent down the following message: We are now quite satisfied as to our general's ability to handle troops. If we are bereft of these two concubines, our meat and drink will lose their savor. It is our wish that they shall not be beheaded.—Sun Tzŭ replied: Having once received His Majesty's commission to be the general of his forces, there are certain commands of His Majesty which, acting in that capacity, I am unable to accept.—Accordingly, he had the two leaders beheaded, and straightway installed the pair next in order as leaders in their place. When this had been done, the drum was sounded for the drill once more; and the girls went through all the evolutions, turning to the right or to the left, marching ahead or wheeling back, kneeling or standing, with perfect accuracy and precision, not venturing to utter a sound. Then Sun Tzŭ sent

> a messenger to the King saying: Your soldiers, Sire, are now properly drilled and disciplined, and ready for your majesty's inspection. They can be put to any use that their sovereign may desire; bid them go through fire and water, and they will not disobey.—But the King replied: Let our general cease drilling and return to camp. As for us, we have no wish to come down and inspect the troops.—Thereupon Sun Tzŭ said: The King is only fond of words, and cannot translate them into deeds.—After that, Ho Lu saw that Sun Tzŭ was one who knew how to handle an army, and finally appointed him general. In the west, he defeated the Ch'u State and forced his way into Ying, the capital; to the north he put fear into the States of Ch'i and Chin, and spread his fame abroad amongst the feudal princes. And Sun Tzŭ shared in the might of the King.

About Sun Tzŭ himself this is all that Ssŭ-ma Ch'ien has to tell us in this chapter. But he proceeds to give a biography of his descendant, Sun Pin, born about a hundred years after his famous ancestor's death, and also the outstanding military genius of his time. The historian speaks of him too as Sun Tzŭ, and in his preface we read: "Sun Tzŭ had his feet cut off and yet continued to discuss the art of war." It seems likely, then, that "Pin" was a nickname bestowed on him after his mutilation, unless the story was invented in order to account for the name. The crowning incident of his career, the crushing defeat of his treacherous rival P'ang Chüan inspired his maxim in [ch. V, sect. 19 of *The Art of War.*]

To return to the elder Sun Tzŭ, he is mentioned in two other passages of the *Shih Chi* [a.k.a. *Records by the Grand Historian* by Ssŭ-ma Ch'ien]:—

> In the third year of his reign [512 B.C.] Ho Lu, king of Wu, took the field with Tzŭ-hsü [*i.e.*, Wu Yüan] and Po P'ei, and attacked Ch'u. He captured the town of Shu and slew the two prince's sons who had formerly been generals of Wu. He was then meditating a descent on Ying [the capital]; but the general Sun Wu said: "The army is exhausted. It is not yet possible. We must wait." . . . [After further successful fighting,] in the ninth year [506 B.C.], King Ho Lu addressed Wu Tzŭ-hsü and Sun Wu, saying: "Formerly, you declared that it was not yet possible for us to enter Ying. Is the time ripe now?" The two men replied: "Ch'u's general Tzŭ-ch'ang, is grasping and covetous, and the princes of T'ang and Ts'ai both have a grudge against him. If Your Majesty has resolved to make a grand attack, you must win over T'ang and Ts'ai, and then you may succeed." Ho Lu followed this advice, [beat Ch'u in five pitched battles and marched into Ying.]

This is the latest date at which anything is recorded of Sun. He does not appear to have survived his patron, who died from the effects of a wound in 496.

[In another chapter] there occurs this passage:

> From this time onward, a number of famous soldiers arose, one after the other: Kao-fan, who was employed by the Chin State; Wang-tzŭ, in the

> service of Ch'i; and Sun Wu, in the service of Wu. These men developed and threw light upon the principles of war.

It is obvious enough that Ssŭ-ma Ch'ien at least had no doubt about the reality of Sun Wu as an historical personage; and with one exception, to be noticed presently, he is by far the most important authority on the period in question. It will not be necessary, therefore, to say much of such a work as the *Wu Yüeh Ch'un Ch'iu,* which is supposed to have been written by Chao Yeh of the 1st century A.D. The attribution is somewhat doubtful; but even if it were otherwise, his account would be of little value, based as it is on the *Shih Chi* and expanded with romantic details. The story of Sun Tzŭ will be found, for what it is worth, in chapter 2. The only new points in it worth noting are: (1) Sun Tzŭ was first recommended to Ho Lu by Wu Tzŭ-hsü. (2) He is called a native of Wu. (3) He had previously lived a retired life, and his contemporaries were unaware of his ability.

The following passage occurs in Huai-nan Tzŭ: "When sovereign and ministers show perversity of mind, it is impossible even for a Sun Tzŭ to encounter the foe." Assuming that this work is genuine (and hitherto no doubt has been cast upon it), we have here the earliest direct reference for Sun Tzŭ, for Huai-nan Tzŭ died in 122 B.C., many years before the *Shih Chi* was given to the world.

Liu Hsiang (80–9 B.C.) in his *Zhan Guo Ce* [*Strategies of the Warring States*] says: "The reason why Sun Tzŭ at the head of 30,000 men beat Ch'u with 200,000 is that the latter were undisciplined."

Têng Ming-shih in his *Ku chin hsing shih shuh pien chu* [*The Study of the Origin of Surnames*] informs us that the surname "Sun" was bestowed on Sun Wu's grandfather by Duke Ching of Ch'i [547–490 B.C.]. Sun Wu's father Sun P'ing, rose to be a Minister of State in Ch'i, and Sun Wu himself, whose style was Ch'ang-ch'ing, fled to Wu on account of the rebellion which was being fomented by the kindred of T'ien Pao. He had three sons, of whom the second, named Ming, was the father of Sun Pin. According to this account then, Pin was the grandson of Wu, which, considering that Sun Pin's victory over Wei was gained in 341 B.C., may be dismissed as chronologically impossible. Whence these data were obtained by Têng Ming-shih I do not know, but of course no reliance whatever can be placed in them.

An interesting document which has survived from the close of the Han period is the short preface written by the Great Ts'ao Ts'ao, or Wei Wu Ti, for his edition of Sun Tzŭ. I shall give it in full:—

> I have heard that the ancients used bows and arrows to their advantage. The *Lun Yü* says: "There must be a sufficiency of military strength." The *Shu Ching* mentions "the army" among the "eight objects of government." The *I Ching* says: "'army' indicates firmness and justice; the experienced leader will have good fortune." The *Shih Ching* says: "The King rose majestic in his wrath, and he marshaled his troops." The Yellow Emperor, T'ang the Completer and Wu Wang all used spears and battle-axes in

order to succor their generation. The *Ssŭ-ma Fa* says: "If one man slay another of set purpose, he himself may rightfully be slain." He who relies solely on warlike measures shall be exterminated; he who relies solely on peaceful measures shall perish. Instances of this are Fu Ch'ai on the one hand and Yen Wang on the other. In military matters, the Sage's rule is normally to keep the peace, and to move his forces only when occasion requires. He will not use armed force unless driven to it by necessity.

Many books have I read on the subject of war and fighting; but the work composed by Sun Wu is the profoundest of them all. [Sun Tzŭ was a native of the Ch'i state, his personal name was Wu. He wrote the Art of War in 13 chapters for Ho Lu, King of Wu. Its principles were tested on women, and he was subsequently made a general. He led an army westwards, crushed the Ch'u state and entered Ying the capital. In the north, he kept Ch'i and Chin in awe. A hundred years and more after his time, Sun Pin lived. He was a descendant of Wu.] In his treatment of deliberation and planning, the importance of rapidity in taking the field, clearness of conception, and depth of design, Sun Tzŭ stands beyond the reach of carping criticism. My contemporaries, however, have failed to grasp the full meaning of his instructions, and while putting into practice the smaller details in which his work abounds, they have overlooked its essential purport. That is the motive which has led me to outline a rough explanation of the whole.

One thing to be noticed in the above is the explicit statement that the 13 chapters were specially composed for King Ho Lu. This is supported by the internal evidence of [ch. I, sect.15], in which it seems clear that some ruler is addressed.

In the bibliographical section of the *Han Shu,* there is an entry which has given rise to much discussion: "The works of Sun Tzŭ of Wu in 82 *p'ien* (or chapters), with diagrams in 9 *chüan.*" It is evident that this cannot be merely the 13 chapters known to Ssŭ-ma Ch'ien, or those we possess today. Chang Shou-chieh in his [*Historical Records*] refers to an edition of Sun Tzŭ's [*The Art of War*] of which the "13 chapters" formed the first *chüan,* adding that there were two other *chüan* besides. This has brought forth a theory, that the bulk of these 82 chapters consisted of other writings of Sun Tzŭ— we should call them apocryphal—similar to the *Wên Ta,* of which a specimen dealing with the Nine Situations [see ch. XI] is preserved in the *T'ung Tien,* and another in Ho Shih's commentary. It is suggested that before his interview with Ho Lu, Sun Tzŭ had only written the 13 chapters, but afterwards composed a sort of exegesis in the form of question and answer between himself and the King. Pí I-hsün, the author of the *Sun Tzŭ Hsü Lu,* backs this up with a quotation from the *Wu Yüeh Ch'un Ch'iu*: "The King of Wu summoned Sun Tzŭ, and asked him questions about the art of war. Each time he set forth a chapter of his work, the King could not find words enough to praise him." As he points out, if the whole work was expounded on the same scale as in the above-mentioned fragments, the total number of chapters could not fail to be considerable. Then the numerous other treatises attributed to Sun Tzŭ might be included. The fact that the *Han Chih* mentions no work of Sun Tzŭ except the 82 *p'ien,* whereas the Sui and T'ang bibliographies give the titles of others in addition to the "13 chapters," is good proof, Pí I-hsün thinks, that all of these were contained in the 82 *p'ien*. Without pinning our faith to the accuracy of details supplied by the *Wu Yüeh Ch'un Ch'iu*, or admitting the genuineness of any of the treatises cited by Pí I-hsün, we may see in this theory a probable solution of the mystery. Between Ssŭ-ma Ch'ien and Pan Ku there was plenty of time for a luxuriant crop of forgeries to have grown up under the magic name of Sun Tzŭ, and the 82 *p'ien* may very well represent a collected edition of these lumped together with the original work. It is also possible, though less likely, that some of them existed in the time of the earlier historian and were purposely ignored by him.

Tu Mu, after Ts'ao Kung the most important commentator on Sun Tzŭ, composed the preface to his edition about the middle of the ninth century. After a somewhat lengthy defence of the military art, he comes at last to Sun Tzŭ himself, and makes two startling assertions:—"The writings of Sun Wu," he said, "originally comprise several hundred thousand words, but Ts'ao Ts'ao, the Emperor Wu Wei, pruned away all redundancies and wrote out the essence of the whole, so as to form a single book in 13 chapters." He goes on to remark that Ts'ao Ts'ao's commentary on Sun Wu leaves a certain proportion of difficulties unexplained. This, in

Tu Mu's opinion, does not necessarily imply that he was unable to furnish a complete commentary. According to the *Wei Chih,* Ts'ao himself wrote a book on war in something over 100,000 words, known as the [*Hsin Shu*]. It appears to have been of such exceptional merit that he suspects Ts'ao to have used for it the surplus material he found in Sun Tzŭ. He concludes, however, by saying "the *Hsin Shu* is now lost, so that the truth cannot be known for certain."

Tu Mu's conjecture seems to be based on a passage [in a book which states:] "Wei Wu Ti strung together Sun Wu's *Art of War,*" which in turn may have resulted from a misunderstanding of the final words of Ts'ao King's preface. This, as Sun Hsing-yen points out, is only a modest way of saying that he made an explanatory paraphrase, or in other words, wrote a commentary on it. On the whole, this theory has met with very little acceptance. Thus, the [*Ssu K'u Ch'uan Shu*] says: "The mention of the 13 chapters in the *Shih Chi* shows that they were in existence before the *Han Chih,* and that latter accretions are not to be considered part of the original work. Tu Mu's assertion can certainly not be taken as proof."

There is every reason to suppose, then, that the 13 chapters existed in the time of Ssŭ-ma Ch'ien practically as we have them now. That the work was then well known he tells us in so many words. "Sun Tzŭ's 13 Chapters and Wu Ch'i's Art of War are the two books that people commonly refer to on the subject of military matters. Both of them are widely distributed, so I will not discuss them here." But as we go further back, serious difficulties begin to arise. The salient fact which has to be faced is that the *Tso Chuan,* the greatest contemporary record, makes no mention whatsoever of Sun Wu, either as a general or as a writer. It is natural, in view of this awkward circumstance, that many scholars should not only cast doubt on the story of Sun Wu as given in the *Shih Chi,* but even show themselves frankly skeptical as to the existence of the man at all. The most powerful presentment of this side of the case is to be found in the following disposition by Yeh Shui-hsin:—

> It is stated in Ssŭ-ma Ch'ien's history that Sun Wu was a native of the Ch'i State, and employed by Wu; and that in the reign of Ho Lu he crushed Ch'u, entered Ying, and was a great general. But in Tso's Commentary no Sun Wu appears at all. It is true that Tso's Commentary need not contain absolutely everything that other histories contain. But Tso has not omitted to mention vulgar plebeians and hireling ruffians such as Ying K'ao-shu, Ts'ao Kuei, Chu Chih-wu and Chüan Shê-chu. In the case of Sun Wu, whose fame and achievements were so brilliant, the omission is much more glaring. Again, details are given, in their due order, about his contemporaries Wu Yüan and the Minister P'ei. Is it credible that Sun Wu alone should have been passed over?
>
> In point of literary style, Sun Tzŭ's work belongs to the same school as *Kuan Tzŭ, Liu T'ao*, and the *Yüeh Yü* and may have been the production of some private scholar living towards the end of the "Spring and Autumn" or the beginning of the

> "Warring States" period. The story that his precepts were actually applied by the Wu State, is merely the outcome of big talk on the part of his followers.
>
> From the flourishing period of the Chou dynasty down to the time of the "Spring and Autumn," all military commanders were statesmen as well, and the class of professional generals, for conducting external campaigns, did not then exist. It was not until the period of the "Six States" that this custom changed. Now although Wu was an uncivilized State, it is conceivable that Tso should have left unrecorded the fact that Sun Wu was a great general and yet held no civil office? What we are told, therefore, about Jang-chü and Sun Wu, is not authentic matter, but the reckless fabrication of theorizing pundits. The story of Ho Lu's experiment on the women, in particular, is utterly preposterous and incredible.

Yeh Shui-hsin represents Ssŭ-ma Ch'ien as having said that Sun Wu crushed Ch'u and entered Ying. This is not quite correct. No doubt the impression left on the reader's mind is that he at least shared in these exploits.... The fact may or may not be significant; but it is nowhere explicitly stated in the *Shih Chi* either that Sun Tzŭ was general on the occasion of the taking of Ying, or that he even went there at all. Moreover, as we know that Wu Yüan and Po P'ei both took part in the expedition, and also that its success was largely due to the dash and enterprise of Fu Kai, Ho Lu's younger brother, it is not easy to see how yet another general could have played a very prominent part in the same campaign.

Ch'ên Chên-sun of the Sung dynasty has the note:—

> Military writers look upon Sun Wu as the father of their art. But the fact that he does not appear in the *Tso Chüan*, although he is said to have served under Ho Lu King of Wu, makes it uncertain what period he really belonged to.

He also says:

> The works of Sun Wu and Wu Ch'i may be of genuine antiquity.

It is noticeable that both Yeh Shui-hsin and Ch'ên Chên-sun, while rejecting the personality of Sun Wu as he figures in Ssŭ-ma Ch'ien's history, are inclined to accept the date traditionally assigned to the work which passes under his name. The author of the Hsü Lu fails to appreciate this distinction, and consequently his bitter attack on Ch'ên Chên-sun really misses its mark. He makes one of two points, however, which certainly tell in favor of the high antiquity of our "13 chapters." "Sun Tzŭ," he says, "must have lived in the age of Ching Wang [519–476], because he is frequently plagiarized in subsequent works of the Chou, Ch'in and Han dynasties." The two most shameless offenders in this respect are Wu Ch'i and Huai-nan Tzŭ, both of them important historical personages in their day. The former lived only a century after the alleged date of Sun Tzŭ, and his death is known to have taken place in 381 B.C. It was to him, according to Liu Hsiang, that Tsêng Shên delivered the *Tso Chuan,* which had been entrusted to him by its author. Now the fact that quotations from the *Art of War,* acknowledged or otherwise, are to be found in so many authors of different epochs, establishes a very strong anterior to them all,— in other words, that Sun Tzŭ's treatise was already in existence towards the end of the 5th century B.C. Further proof of Sun Tzŭ's antiquity is furnished by the archaic or wholly obsolete meanings attaching to a number of the words he uses. A list of these, which might perhaps be extended, is given in the *Hsü Lu*; and though some of the interpretations are doubtful, the main argument is hardly affected thereby. Again, it must not be forgotten that Yeh Shui-hsin, a scholar and critic of the first rank, deliberately pronounces the style of the 13 chapters to belong to the early part of the fifth century. Seeing that he is actually engaged in an attempt to disprove the existence of Sun Wu himself, we may be sure that he would not have hesitated to assign the work to a later date had he not honestly believed the contrary. And it is precisely on such a point that the judgment of an educated Chinaman will carry most weight. Other internal evidence is not far to seek. Thus in ch. XIII, sect. 1, there is an unmistakable allusion to the ancient system of land-tenure which had already passed away by the time of Mencius, who was anxious to see it revived in a modified form. The only warfare Sun Tzŭ knows is that carried on between the various feudal princes, in which armored chariots play a large part. Their use seems to have entirely died out before the end of the Chou dynasty. He speaks as a man of Wu, a state which ceased to exist as early as 473 B.C. On this I shall touch presently.

But once refer the work to the 5th century or earlier, and the chances of its being other than a *bonâ fide* production are sensibly diminished. The great age of forgeries did not come until long after. That it should have been forged in the period immediately following 473 is particularly unlikely, for no one, as a rule, hastens to identify himself with a lost cause. As for Yeh Shui-hsin's theory, that the author was a literary recluse, that seems to me quite untenable. If one thing is more apparent than another after reading the maxims of Sun Tzŭ, it is that their essence has been distilled from a large store of personal observation and experience. They reflect the mind not only of a born strategist, gifted with a rare faculty of generalization, but also of a practical soldier closely acquainted with the military conditions of his time. To say nothing of the fact that these sayings have been accepted and endorsed by all the greatest captains of Chinese history, they offer a combination of freshness and sincerity, acuteness and common sense, which quite excludes the idea that they were artificially concocted in the study. If we admit, then, that the 13 chapters were the genuine production of a military man living towards the end of the "Ch'un Ch'iu" period, are we not bound, in spite of the silence of the *Tso Chuan,* to accept Ssŭ-ma Ch'ien's account in its entirety? In view of his high repute as a sober historian, must we not hesitate to assume that the records he drew upon for Sun Wu's biography were false and untrustworthy? The answer, I fear, must be in the negative. There is still one grave, if not fatal, objection to the chronology involved in the story as told in the *Shih Chi,* which, so far as I am aware, nobody has yet pointed out. There are two passages in Sun Tzŭ in which he alludes to contemporary affairs. The first is in ch. VI, sect. 21:—

> Though according to my estimate the soldiers of Yüeh exceed our own in number, that shall advantage them nothing in the matter of victory. I say then that victory can be achieved.

The other is in ch. XI, sect. 30:

> Asked if an army can be made to imitate the *shuai-jan,* I should answer, Yes. For the men of Wu and the men of Yüeh are enemies; yet if they are crossing a river in the same boat and are caught by a storm, they will come to each other's assistance just as the left hand helps the right.

These two paragraphs are extremely valuable as evidence of the date of composition. They assign the work to the period of the struggle between Wu and Yüeh. So much has been observed by Pí I-hsün. But what has hitherto escaped notice is that they also seriously impair the credibility of Ssŭ-ma Ch'ien's narrative. As we have seen above, the first positive date given in connection with Sun Wu is 512 B.C. He is then spoken of as a general, acting as confidential adviser to Ho Lu, so that his alleged introduction to that monarch had already taken place, and of course the 13 chapters must have been written earlier still. But at that time, and for several years after, down to the capture of Ying in 506, Ch'u and not Yüeh, was the great hereditary enemy of Wu. The two states, Ch'u and Wu, had been constantly at war for over half a century, whereas the first war between Wu and Yüeh was waged only in 510, and even then was no more than a short interlude sandwiched in the midst of the fierce struggle with Ch'u. Now Ch'u is not mentioned in the 13 chapters at all. The natural inference is that they were written at a time when Yüeh had become the prime antagonist of Wu, that is, after Ch'u had suffered the great humiliation of 506. At this point, a table of dates may be found useful.

B.C.	
514	Accession of Ho Lu.
512	Ho Lu attacks Ch'u, but is dissuaded from entering Ying, the capital. *Shi Chi* mentions Sun Wu as general.
511	Another attack on Ch'u.
510	Wu makes a successful attack on Yüeh. This is the first war between the two states.
509 or 508	Ch'u invades Wu, but is signally defeated at Yü-chang.
506	Ho Lu attacks Ch'u with the aid of T'ang and Ts'ai. Decisive battle of Po-chü, and capture of Ying. Last mention of Sun Wu in *Shih Chi*.
505	Yüeh makes a raid on Wu in the absence of its army. Wu is beaten by Ch'in and evacuates Ying.
504	Ho Lu sends Fu Ch'ai to attack Ch'u.
497	Kou Chien becomes King of Yüeh.

496	Wu attacks Yüeh, but is defeated by Kou Chien at Tsui-li. Ho Lu is killed.
494	Fu Ch'ai defeats Kou Chien in the great battle of Fu-chiao, and enters the capital of Yüeh.
485 or 484	Kou Chien renders homage to Wu. Death of Wu Tzŭ-hsü.
482	Kou Chien invades Wu in the absence of Fu Ch'ai.
478 to 476	Further attacks by Yüeh on Wu.
475	Kou Chien lays siege to the capital of Wu.
473	Final defeat and extinction of Wu.

The sentence quoted above from ch. VI, sect. 21 hardly strikes me as one that could have been written in the full flush of victory. It seems rather to imply that, for the moment at least, the tide had turned against Wu, and that she was getting the worst of the struggle. Hence we may conclude that our treatise was not in existence in 505, before which date Yüeh does not appear to have scored any notable success against Wu. Ho Lu died in 496, so that if the book was written for him, it must have been during the period 505–496, when there was a lull in the hostilities, Wu having presumably exhausted by its supreme effort against Ch'u. On the other hand, if we choose to disregard the tradition connecting Sun Wu's name with Ho Lu, it might equally well have seen the light between 496 and 494, or possibly in the period 482–473, when Yüeh was once again becoming a very serious menace. We may feel fairly certain that the author, whoever he may have been, was not a man of any great eminence in his own day. On this point the negative testimony of the *Tso Tchuan* far outweighs any shred of authority still attaching to the *Shih Chi,* if once its other facts are discredited. Sun Hsing-yen, however, makes a feeble attempt to explain the omission of his name from the great commentary. It was Wu Tzŭ-hsü, he says, who got all the credit of Sun Wu's exploits, because the latter (being an alien) was not rewarded with an office in the State.

How then did the Sun Tzŭ legend originate? It may be that the growing celebrity of the book imparted by degrees a kind of factitious renown to its author. It was felt to be only right and proper that one so well versed in the science of war should have solid achievements to his credit as well. Now the capture of Ying was undoubtedly the greatest feat of arms in Ho Lu's reign; it made a deep and lasting impression on all the surrounding states, and raised Wu to the short-lived zenith of her power. Hence, what more natural, as time went on, than that the acknowledged master of strategy, Sun Wu, should be popularly identified with that campaign, at first perhaps only in the sense that his brain conceived and planned it; afterwards, that it was actually carried out by him in conjunction with Wu Yüan, Po P'ei and Fu Kai?

It is obvious that any attempt to reconstruct even the outline of Sun Tzŭ's life must be based almost wholly on conjecture. With this necessary proviso, I should say that he probably entered the service of Wu about the time of Ho Lu's accession, and gathered experience, though only in the capacity of a subordinate officer, during the intense military activity which marked the first half of the prince's reign. If he rose to be a general at all, he certainly was never on an equal footing with the three above mentioned. He was doubtless present at the investment and occupation of Ying, and witnessed Wu's sudden collapse in the following year. Yüeh's attack at this critical juncture, when her rival was embarrassed on every side, seems to have convinced him that this upstart kingdom was the great enemy against whom every effort would henceforth have to be directed. Sun Wu was thus a well-seasoned warrior when he sat down to write his famous book, which according to my reckoning must have appeared towards the end, rather than the beginning of Ho Lu's reign. The story of the women may possibly have grown out of some real incident occurring about the same time. As we hear no more of Sun Wu after this from any source, he is hardly likely to have survived his patron or to have taken part in the death-struggle with Yüeh, which began with the disaster at Tsui-li.

If these inferences are approximately correct, there is a certain irony in the fate which decreed that China's most illustrious man of peace should be contemporary with her greatest writer on war.

THE TEXT OF SUN TZŬ

I have found it difficult to glean much about the history of Sun Tzŭ's text. The quotations that occur in early authors go to show that the "13 chapters" of which Ssŭ-ma Ch'ien speaks were essentially the same as those now extant. We have his word for it that they were widely circulated in his day, and can only regret that he refrained from discussing them on that account. Sun Hsing-yen says in his preface:—

> During the Ch'in and Han dynasties Sun Tzŭ's *Art of War* was in general use amongst military commanders, but they seem to have treated it as a work of mysterious import, and were unwilling to expound it for the benefit of posterity. Thus it came about that Wei Wu was the first to write a commentary on it.

As we have already seen, there is no reasonable ground to suppose that Ts'ao Kung tampered with the text. But the text itself is often so obscure, and the number of editions which appeared from that time onward so great, especially during the T'ang and Sung dynasties, that it would be surprising if numerous corruptions had not managed to creep in. Towards the middle of the Sung period, by which time all the chief commentaries on Sun Tzŭ were in existence, a certain Chi T'ien-pao published a work in 15 *chüan* entitled "Sun Tzŭ with the collected commentaries of ten writers." There was another text, with variant readings put forward by Chu Fu of Ta-hsing, which also had supporters among the scholars of that period; but in the Ming editions, Sun Hsing-yen tells us, these readings were for some reason or other no longer put into circulation. Thus, until the end of the 18th century, the text in sole possession of the field was one derived from Chi T'ien-pao's edition, although no actual copy of that important work was known to have survived. That, therefore, is the text of Sun Tzŭ which appears in the War section of the great Imperial encyclopedia printed in 1726, the *Ku Chin T'u Shu Chi Ch'êng.* Another copy at my disposal of what is practically the same text, with slight variations, is that contained in the "Eleven philosophers of the Chou and Ch'in dynasties" (1758). And the Chinese printed in Captain Calthrop's first edition is evidently a similar version which has filtered through Japanese channels. So things remained until Sun Hsing-yen (1752–1818), a distinguished antiquarian and classical scholar, who claimed to be an actual descendant of Sun Wu, accidentally discovered a copy of Chi T'ien-pao's long-lost work, when on a visit to the library of the Hua-yin temple. Appended to it was the *I Shuo* of Chêng Yu-hsien, mentioned in the *T'ung Chih,* and also believed to have perished. This is what Sun Hsing-yen designates as the . . . "original edition (or text)"—a rather misleading name, for it cannot by any means claim to set before us the text of Sun Tzŭ in its pristine purity. Chi T'ien-pao was a careless compiler, and appears to have been content to reproduce the somewhat debased version current in his day, without troubling to collate it with the earliest editions then available. Fortunately, two versions of Sun Tzŭ, even older than the newly discovered work, were still extant, one buried in the *T'ung Tien,* Tu Yu's great treatise on the Constitution, the other similarly enshrined in the *T'ai P'ing Yü Lan* encyclopedia. In both the complete text is to be found, though split up into fragments, intermixed with other matter, and scattered piecemeal over a number of different sections. Considering that the *Yü Lan* takes us back to the year 983, and the *T'ung Tien* about 200 years further still, to the middle of the T'ang dynasty, the value of these early transcripts of Sun Tzŭ can hardly be overestimated. Yet the idea of utilizing them does not seem to have occurred to anyone until Sun Hsing-yen, acting under Government instructions, undertook a thorough recension of the text. This is his own account:—

> Because of the numerous mistakes in the text of Sun Tzŭ which his editors had handed down, the Government ordered that the ancient edition [of Chi T'ien-pao] should be used, and that the text should be revised and corrected throughout. It happened that Wu Nien-hu, the Governor Pi Kua, and Hsi, a graduate of the second degree, had all devoted themselves to this study, probably surpassing me therein. Accordingly, I have had the whole work cut on blocks as a textbook for military men.

The three individuals here referred to had evidently been occupied on the text of Sun Tzŭ prior to Sun Hsing-yen's commission, but we are left in doubt as to the work they really accomplished. At any rate, the new edition, when ultimately produced, appeared in the names of Sun Hsing-yen and only one co-editor, Wu Jên-chi. They took the "original text" as their basis, and by careful comparison with older versions, as well as the extant commentaries and other sources of information

> such as the *I Shuo,* succeeded in restoring a very large number of doubtful passages, and turned out, on the whole, what must be accepted as the closest approximation we are ever likely to get to Sun Tzŭ's original work. This is what will hereafter be denominated the "standard text."

The copy which I have used belongs to a reissue dated 1877. It is in 6 *pên*, forming part of a well-printed set of 23 early philosophical works in 83 *pên*. It opens with a preface by Sun Hsing-yen (largely quoted in this introduction), vindicating the traditional view of Sun Tzŭ's life and performances, and summing up in remarkably concise fashion the evidence in its favor. This is followed by Ts'ao Kung's preface to his edition, and the biography of Sun Tzŭ from the *Shih Chi,* both translated above. Then come, firstly, Chêng Yu-hsien's *I Shuo,* with author's preface, and next, a short miscellany of historical and bibliographical information entitled *Sun Tzŭ Hsü Lu,* compiled by Pi I-hsün. As regards the body of the work, each separate sentence is followed by a note on the text, if required, and then by the various commentaries appertaining to it, arranged in chronological order. These we shall now proceed to discuss briefly, one by one.

THE COMMENTATORS

Sun Tzŭ can boast an exceptionally long distinguished roll of commentators, which would do honor to any classic. Ou-yang Hsiu remarks on this fact, though he wrote before the tale was complete, and rather ingeniously explains it by saying that the artifices of war, being inexhaustible, must therefore be susceptible of treatment in a great variety of ways.

Ts'ao Ts'ao or Ts'ao Kung, afterwards known as Wei Wu Ti (A.D. 155–220). There is hardly any room for doubt that the earliest commentary on Sun Tzŭ actually came from the pen of this extraordinary man, whose biography in the *San Kuo Chih* reads like a romance. One of the greatest military geniuses that the world has seen, and Napoleonic in the scale of his operations, he was especially famed for the marvellous rapidity of his marches, which has found expression in the line "Talk of Ts'ao Ts'ao, and Ts'ao Ts'ao will appear." Ou-yang Hsiu says of him that he was a great captain who "measured his strength against Tung Cho, Lü Pu and the two Yüan, father and son, and vanquished them all; whereupon he divided the Empire of Han with Wu and Shu, and made himself king. It is recorded that whenever a council of war was held by Wei on the eve of a far-reaching campaign, he had all his calculations ready; those generals who made use of them did not lose one battle in ten; those who ran counter to them in any particular saw their armies incontinently beaten and put to flight." Ts'ao Kung's notes on Sun Tzŭ, models of austere brevity, are so thoroughly characteristic of the stern commander known to history, that it is hard indeed to conceive of them as the work of a mere *littérateur*. Sometimes, indeed, owing to extreme compression, they are scarcely intelligible and stand no less in need of a commentary than the text itself.

Mêng Shih. The commentary which has come down to us under this name is comparatively meager, and nothing about the author is known. Even his personal name has not been recorded. Chi T'ien-pao's edition places him after Chia Lin, and Ch'ao Kung-wu also assigns him to the T'ang dynasty, but this is a mistake. In Sun Hsing-yen's preface, he appears as Mêng Shih of the Liang dynasty (502–557). Others would identify him with Mêng K'ang of the 3rd century. [In one book] he is named last of the "Five Commentators," the others being Wei Wu Ti, Tu Mu, Ch'ên Hao and Chia Lin.

Li Ch'üan of the 8th century was a well-known writer on military tactics. . . . [One work] mentions [a book on the] lives of famous generals from the Chou to the T'ang dynasty as written by him. According to Ch'ao Kung-wu and the *T'ien-i-ko* catalogue, he followed [a variant] text of Sun Tzŭ which differs considerably from those now extant. His notes are mostly short and to the point, and he frequently illustrates his remarks by anecdotes from Chinese history.

Tu Yu (died 812) did not publish a separate commentary on Sun Tzŭ, his notes being taken from the *T'ung Tien,* the encyclopaedic treatise on the Constitution which was his life-work. They are largely repetitions of Ts'ao Kung and Mêng Shih, besides which it is believed that he drew on the ancient commentaries of Wang Ling and others. Owing to the peculiar arrangement of *T'ung Tien,* he has to explain each passage on its merits, apart from the context, and sometimes his own explanation does not agree with that of Ts'ao Kung, whom he always quotes first. Though not strictly to be reckoned as one of the "Ten Commentators," he was added to their number by Chi T'ien-pao, being wrongly placed after his grandson Tu Mu.

Tu Mu (803-852) is perhaps best known as a poet—a bright star even in the glorious galaxy of the T'ang period. We learn from Ch'ao Kung-wu that although he had no practical experience of war, he was extremely fond of discussing the subject, and was moreover well read in the military history of the *Ch'un Ch'iu* and *Chan Kuo* eras. His notes, therefore, are well worth attention. They are very copious, and replete with historical parallels. The gist of Sun Tzŭ's work is thus summarized by him: "Practise benevolence and justice, but on the other hand make full use of artifice and measures of expediency." He further declared that all the military triumphs and disasters of the thousand years which had elapsed since Sun Tzŭ's death would, upon examination, be found to uphold and corroborate, in every particular, the maxims contained in his book. Tu Mu's somewhat spiteful charge against Ts'ao Kung has already been considered elsewhere.

Ch'ên Hao appears to have been a contemporary of Tu Mu. Ch'ao Kung-wu says that he was impelled to write a new commentary on Sun Tzŭ because Ts'ao Kung's on the one hand was too obscure and subtle, and that of Tu Mu on the other too long-winded and diffuse.

Ou-yang Hsiu, writing in the middle of the 11th century, calls Ts'ao Kung, Tu Mu and Ch'ên Hao the three chief commentators on Sun Tzŭ, and observes that Ch'ên Hao is continually attacking Tu Mu's shortcomings. His commentary, though not lacking in merit, must rank below those of his predecessors.

Chia Lin is known to have lived under the T'ang dynasty, for his commentary on Sun Tzŭ is mentioned in the T'ang Shu and was afterwards republished by Chi Hsieh of the same dynasty together with those of Mêng Shih and Tu Yu. It is of somewhat scanty texture, and in point of quality, too, perhaps the least valuable of the eleven.

Mei Yao-ch'ên (1002–1060), commonly known by his "style" as Mei Shêng-yü, was, like Tu Mu, a poet of distinction. His commentary was published with a laudatory preface by the great Ou-yang Hsiu, from which we may cull the following:—

> Later scholars have misread Sun Tzŭ, distorting his words and trying to make them square with their own one-sided views. Thus, though commentators have not been lacking, only a few have proved equal to the task. My friend Shêng-yü has not fallen into this mistake. In attempting to provide a critical commentary for Sun Tzŭ's work, he does not lose sight of the fact that these sayings were intended for states engaged in internecine warfare; that the author is not concerned with the military conditions prevailing under the sovereigns of the three ancient dynasties, nor with the

> nine punitive measures prescribed to the Minister of War. Again, Sun Wu loved brevity of diction, but his meaning is always deep. Whether the subject be marching an army, or handling soldiers, or estimating the enemy, or controlling the forces of victory, it is always systematically treated; the sayings are bound together in strict logical sequence, though this has been obscured by commentators who have probably failed to grasp their meaning. In his own commentary, Mei Shêng-yü has brushed aside all the obstinate prejudices of these critics, and has tried to bring out the true meaning of Sun Tzŭ himself. In this way, the clouds of confusion have been dispersed and the sayings made clear. I am convinced that the present work deserves to be handed down side by side with the three great commentaries; and for a great deal that they find in the sayings, coming generations will have constant reason to thank my friend Shêng-yü.

Making some allowance for the exuberance of friendship, I am inclined to endorse this favorable judgment, and would certainly place him above Ch'ên Hao in order of merit.

Wang Hsi, also of the Sung dynasty, is decidedly original in some of his interpretations, but much less judicious than Mei Yao-ch'ên, and on the whole not a very trustworthy guide. He is fond of comparing his own commentary with that of Ts'ao Kung, but the comparison is not often flattering to him. We learn from Ch'ao Kung-wu that Wang Hsi revised the ancient text of Sun Tzŭ, filling up lacunae and correcting mistakes.

Ho Yen-hsi of the Sung dynasty. The personal name of this commentator is given as above by Chêng Ch'iao in the *T'ung Chih,* written about the middle of the twelfth century, but he appears simply as Ho Shih in the *Yu Hai,* and Ma Tuan-lin quotes Ch'ao Kung-wu as saying that his personal name is unknown. There seems to be no reason to doubt Chêng Ch'iao's statement, otherwise I should have been inclined to hazard a guess and identify him with one Ho Ch'ü-fei, the author of a short treatise on war . . . who lived in the latter part of the 11th century. Ho Shih's commentary, in the words of the *T'ien-i-Ko* catalogue, "contains helpful additions" here and there, but is chiefly remarkable for the copious extracts taken, in adapted form, from the dynastic histories and other sources.

Chang Yü. The list closes with a commentator of no great originality perhaps, but gifted with admirable powers of lucid exposition. His commentator is based on that of Ts'ao Kung, whose terse sentences he contrives to expand and develop in masterly fashion. Without Chang Yü, it is safe to say that much of Ts'ao Kung's commentary would have remained cloaked in its pristine obscurity and therefore valueless. His work is not mentioned in the Sung history, the *T'ung K'ao,* or the *Yü Hai,* but it finds a niche in the *T'ung Chih,* which also names him as the author of the "Lives of Famous Generals."

It is rather remarkable that the last-named four should all have flourished within so short a space of time. Ch'ao Kung-wu accounts for it by saying: "During the early years of the Sung dynasty the Empire enjoyed a long spell of peace, and men ceased to practice the art of war. But when (Chao) Yüan-hao's rebellion came [1038–1042] and the frontier generals were defeated time after time, the Court made strenuous inquiry for men skilled in war, and military topics became the vogue amongst all the high officials. Hence it is that the commentators of Sun Tzŭ in our dynasty belong mainly to that period."

Besides these eleven commentators, there are several others whose work has not come down to us. The *Sui Shu* mentions four, namely Wang Ling (often quoted by Tu Yü as Wang Tzŭ); Chang Tzŭ- shang; Chia Hsü of Wei; and Shên Yu of Wu. The *T'ang Shu* adds Sun Hao, and the *T'ung Chih* Hsiao Chi, while the *T'u Shu* mentions a Ming commentator, Huang Jun-yu. It is possible that some of these may have been merely collectors and editors of other commentaries, like Chi T'ien-pao and Chi Hsieh, mentioned above. Certainly in the case of the latter, the entry in the *T'ung K'ao,* without the following note, would give one to understand that he had written an independent commentary of his own.

There are two works, described in the *Ssu K'u Ch'üan Shu* and no doubt extremely rare, which I should much like to have seen. [One appears] in 5 *chüan*. It gives selections from four new commentators, probably of the Ming dynasty, as well as from the eleven known to us. The names of the four are Hsieh Yüan; Chang Ao; Li Ts'ai; and Huang Chih-chêng. The other work is in 4 *chüan*, compiled by Chêng Tuan of the present dynasty. It is a compendium of information on Sun Tzŭ's 13 chapters.

APPRECIATIONS OF SUN TZŬ

Sun Tzŭ has exercised a potent fascination over the minds of some of China's greatest men. Among the famous generals who are known to have studied his pages with enthusiasm may be mentioned Han Hsin (d. 196 B.C.), Fêng I (d. A.D. 34), Lü Mêng (d. 219), and Yo Fei (1103–1141). The opinion of Ts'ao Kung, who disputes with Han Hsin the highest place in Chinese military annals, has already been recorded. Still more remarkable, in one way, is the testimony of purely literary men, such as Su Hsün (the father of Su Tung-p'o), who wrote several essays on military topics, all of which owe their chief inspiration to Sun Tzŭ. The following short passage by him is preserved in the *Yü Hai*:—

> Sun Wu's saying, that in war one cannot make certain of conquering, is very different indeed from what other books tell us. Wu Ch'i was a man of the same stamp as Sun Wu: they both wrote books on war, and they are linked together in popular speech as "Sun and Wu." But Wu Ch'i's remarks on war are less weighty, his rules are rougher and more crudely stated, and there is not the same unity of plan as in Sun Tzŭ's work, where the style is terse, but the meaning fully brought out.

[Chapter 17 of the "Impartial Judgments in the Garden of Literature" by Chêng Hou contains the following extract]:

> Sun Tzŭ's 13 chapters are not only the staple and base of all military men's training, but also compel the most careful attention of scholars and men of letters. His sayings are terse yet elegant, simple yet profound, perspicuous and eminently practical. Such works as the *Lun Yü,* the *I Ching* and the great Commentary, as well as the writings of Mencius, Hsün K'uang and Yang Chu, all fall below the level of Sun Tzŭ.

Chu Hsi, commenting on this, fully admits the first part of the criticism, although he dislikes the audacious comparison with the venerated classical works. Language of this sort, he says, "encourages a ruler's bent towards unrelenting warfare and reckless militarism."

APOLOGIES FOR WAR

Accustomed as we are to think of China as the greatest peace-loving nation on earth, we are in some danger of forgetting that her experience of war in all its phases has also been such as no modern State can parallel. Her long military annals stretch back to a point at which they are lost in the mists of time. She had built the Great Wall and was maintaining a huge standing army along her frontier centuries before the first Roman legionary was seen on the Danube. What with the perpetual collisions of the ancient feudal States, the grim conflicts with Huns, Turks and other invaders after the centralization of government, the terrific upheavals which accompanied the overthrow of so many dynasties, besides the countless rebellions and minor disturbances that have flamed up and flickered out again one by one, it is hardly too much to say that the clash of arms has never ceased to resound in one portion or another of the Empire.

No less remarkable is the succession of illustrious captains to whom China can point with pride. As in all countries, the greatest are fond of emerging at the most fateful crises of her history. Thus, Po Ch'i stands out conspicuous in the period when Ch'in was entering upon her final struggle with the remaining independent states. The stormy years which followed the break-up of the Ch'in dynasty are illuminated by the transcendent genius of Han Hsin. When the House of Han in turn is tottering to its fall, the great and baleful figure of Ts'ao Ts'ao dominates the scene. And in the establishment of the T'ang dynasty, one of the mightiest tasks achieved by man, the superhuman energy of Li Shih-min (afterwards the Emperor T'ai Tsung) was seconded by the brilliant strategy of Li Ching. None of these generals need fear comparison with the greatest names in the military history of Europe.

In spite of all this, the great body of Chinese sentiment, from Lao Tzŭ downwards, and especially as reflected in the standard literature of Confucianism, has been consistently pacific and intensely opposed to militarism in any form. It is such an uncommon thing to find any of the literati defending warfare on principle, that I have thought it worth while to collect and translate a few passages in which the unorthodox view is upheld. The following, by Ssŭ-ma Ch'ien, shows that for all his ardent admiration of Confucius, he was yet no advocate of peace at any price:

> Military weapons are the means used by the Sage to punish violence and cruelty, to give peace to troublous times, to remove difficulties and dangers, and to succor those who are in peril. Every animal with blood in its veins and horns on its head will fight when it is attacked. How much more so will man, who carries in his breast the faculties of love and hatred, joy and anger! When he is pleased, a feeling of affection springs up within him; when angry, his poisoned sting is brought into play. That is the natural law which governs his being. . . . What then shall be said of those scholars of our time, blind to all great issues, and without any appreciation of relative values, who can only bark out their stale formulas about "virtue" and "civilization," condemning the use of military weapons? They will surely bring our country to impotence and dishonor and the loss of her rightful heritage; or, at the very least, they will bring about invasion and rebellion, sacrifice of territory and general enfeeblement. Yet they obstinately refuse to modify the position they have taken up. The truth is that, just as in the family the teacher must not spare the rod, and punishments cannot be dispensed with in the State, so military chastisement can never be allowed to fall into abeyance in the Empire. All one can say is that this power will be exercised wisely by some, foolishly by others, and that among those who bear arms some will be loyal and others rebellious.

The next piece is taken from Tu Mu's preface to his commentary on Sun Tzŭ:

War may be defined as punishment, which is one of the functions of government. It was the profession of Chung Yu and Jan Ch'iu, both disciples of Confucius. Nowadays, the holding of trials and hearing of litigation, the imprisonment of offenders and their execution by flogging in the market-place, are all done by officials. But the wielding of huge armies, the throwing down of fortified cities, the hauling of women and children into captivity, and the beheading of traitors—this is also work which is done by officials. The objects of the rack and of military weapons are essentially the same. There is no intrinsic difference between the punishment of flogging and cutting off heads in war. For the lesser infractions of law, which are easily dealt with, only a small amount of force need be employed: hence the use of military weapons and wholesale decapitation. In both cases, however, the end in view is to get rid of wicked people, and to give comfort and relief to the good.

Chi-sun asked Jan Yu, saying: "Have you, Sir, acquired your military aptitude by study, or is it innate?" Jan Yu replied: "It has been acquired by study." "How can that be so," said Chi-sun, "seeing that you are a disciple of Confucius?" "It is a fact," replied Jan Yu; "I was taught by Confucius. It is fitting that the great Sage should exercise both civil and military functions, though to be sure my instruction in the art of fighting has not yet gone very far."

> Now, who the author was of this rigid distinction between the "civil" and the "military," and the limitation of each to a separate sphere of action, or in what year of which dynasty it was first introduced, is more than I can say. But, at any rate, it has come about that the members of the governing class are quite afraid of enlarging on military topics, or do so only in a shamefaced manner. If any are bold enough to discuss the subject, they are at once set down as eccentric individuals of coarse and brutal propensities. This is an extraordinary instance in which, through sheer lack of reasoning, men unhappily lose sight of fundamental principles.
>
> When the Duke of Chou was minister under Ch'êng Wang, he regulated ceremonies and made music, and venerated the arts of scholarship and learning; yet when the barbarians of the River Huai revolted, he sallied forth and chastised them. When Confucius held office under the Duke of Lu, and a meeting was convened at Chia-ku, he said: "If pacific negotiations are in progress, warlike preparations should have been made beforehand." He rebuked and shamed the Marquis of Ch'i, who cowered under him and dared not proceed to violence. How can it be said that these two great Sages had no knowledge of military matters?

We have seen that the great Chu Hsi held Sun Tzŭ in high esteem. He also appeals to the authority of the Classics:—

> Our Master Confucius, answering Duke Ling of Wei, said: "I have never studied matters connected with armies and battalions." Replying to K'ung Wên-tzŭ, he said "I have not been instructed about buff-coats and weapons." But if we turn to the meeting at Chia-ku, we find that he used armed force against the men of Lai, so that the marquis of Ch'i was overawed. Again, when the inhabitants of Pi revolted, he ordered his officers to attack them, whereupon they were defeated and fled in confusion. He once uttered the words: "If I fight, I conquer." And Jan Yu also said: "The Sage exercises both civil and military functions." Can it be a fact that Confucius never studied or received instruction in the art of war? We can only say that he did not specially choose matters connected with armies and fighting to be the subject of his teaching.

Sun Hsing-yen, the editor of Sun Tzŭ, writes in similar strain:—

> Confucius said: "I am unversed in military matters." He also said: "If I fight, I conquer." Confucius ordered ceremonies and regulated music. Now war constitutes one of the five classes of State ceremonial, and must not be treated as an independent branch of study. Hence, the words "I am unversed in" must be taken to mean that there are things which even an inspired Teacher does not know. Those who have to lead an army and devise stratagems, must learn the art of war. But if one can

> command the services of a good general like Sun Tzŭ, who was employed by Wu Tzŭ-hsü, there is no need to learn it oneself. Hence the remark added by Confucius: "If I fight, I conquer."

The men of the present day, however, wilfully interpret these words of Confucius in their narrowest sense, as though he meant that books on the art of war were not worth reading. With blind persistency, they adduce the example of Chao Kua, who pored over his father's books to no purpose, as a proof that all military theory is useless. Again, seeing that books on war have to do with such things as opportunism in designing plans, and the conversion of spies, they hold that the art is immoral and unworthy of a sage. These people ignore the fact that the studies of our scholars and the civil administration of our officials also require steady application and practice before efficiency is reached. The ancients were particularly chary of allowing mere novices to botch their work. Weapons are baneful and fighting perilous; and useless unless a general is in constant practice, he ought not to hazard other men's lives in battle. Hence it is essential that Sun Tzŭ's 13 chapters should be studied. Hsiang Liang used to instruct his nephew Chi in the art of war. Chi got a rough idea of the art in its general bearings, but would not pursue his studies to their proper outcome, the consequence being that he was finally defeated and overthrown. He did not realize that the tricks and artifices of war are beyond verbal computation. Duke Hsiang of Sung and King Yen of

Hsü were brought to destruction by their misplaced humanity. The treacherous and underhand nature of war necessitates the use of guile and stratagem suited to the occasion. There is a case on record of Confucius himself having violated an extorted oath, and also of his having left the Sung State in disguise. Can we then recklessly arraign Sun Tzŭ for disregarding truth and honesty?

1

LAYING PLANS

1. Sun Tzŭ said: The art of war is of vital importance to the State.

2. It is a matter of life and death, a road either to safety or to ruin. Hence it is a subject of inquiry which can on no account be neglected.

3. The art of war, then, is governed by five constant factors, to be taken into account in one's deliberations, when seeking to determine the conditions obtaining in the field.

4. These are: (1) The Moral Law; (2) Heaven; (3) Earth; (4) The Commander; (5) Method and discipline.

5, 6. *The Moral Law* causes the people to be in complete accord with their ruler, so that they will follow him regardless of their lives, undismayed by any danger.

7. *Heaven* signifies night and day, cold and heat, times and seasons.

8. *Earth* comprises distances, great and small; danger and security; open ground and narrow passes; the chances of life and death.

9. *The Commander* stands for the virtues of wisdom, sincerity, benevolence, courage and strictness.

10. By *Method and discipline* are to be understood the marshalling of the army in its proper subdivisions, the graduations of rank among the officers, the maintenance of roads by which supplies may reach the army, and the control of military expenditure.

11. These five heads should be familiar to every general: he who knows them will be victorious; he who knows them not will fail.

12. Therefore, in your deliberations, when seeking to determine the military conditions, let them be made the basis of a comparison, in this wise:—

13. (1) Which of the two sovereigns is imbued with the Moral law?

(2) Which of the two generals has most ability?

(3) With whom lie the advantages derived from Heaven and Earth?

(4) On which side is discipline most rigorously enforced?[1]

(5) Which army is stronger?

(6) On which side are officers and men more highly trained?

(7) In which army is there the greater constancy both in reward and punishment?

14. By means of these seven considerations I can forecast victory or defeat.

15. The general that hearkens to my counsel and acts upon it, will conquer:—let such a one be retained in command! The general that hearkens not to my counsel nor acts upon it, will suffer defeat:—let such a one be dismissed!

16. While heading the profit of my counsel, avail yourself also of any helpful circumstances over and beyond the ordinary rules.

17. According as circumstances are favorable, one should modify one's plans.

18. All warfare is based on deception.

19. Hence, when able to attack, we must seem unable; when using our forces, we must seem inactive; when we are near, we must make the enemy believe we are far away; when far away, we must make him believe we are near.

20. Hold out baits to entice the enemy. Feign disorder, and crush him.

21. If he is secure at all points, be prepared for him. If he is in superior strength, evade him.

22. If your opponent is of choleric temper, seek to irritate him. Pretend to be weak, that he may grow arrogant.

23. If he is taking his ease, give him no rest. If his forces are united, separate them.

24. Attack him where he is unprepared, appear where you are not expected.

25. These military devices, leading to victory, must not be divulged beforehand.

26. Now the general who wins a battle makes many calculations in his temple ere the battle is fought. The general who loses a battle makes but few calculations beforehand. Thus do many calculations lead to victory, and few calculations to defeat: how much more no calculation at all! It is by attention to this point that I can foresee who is likely to win or lose.

II

WAGING WAR

1. Sun Tzŭ said: In the operations of war, where there are in the field a thousand swift chariots, as many heavy chariots, and a hundred thousand mail-clad soldiers, with provisions enough to carry them a thousand *li,* the expenditure at home and at the front, including entertainment of guests, small items such as glue and paint, and sums spent on chariots and armour, will reach the total of a thousand ounces of silver per day. Such is the cost of raising an army of 100,000 men.

2. When you engage in actual fighting, if victory is long in coming, then men's weapons will grow dull and their ardour will be damped. If you lay siege to a town, you will exhaust your strength.

3. Again, if the campaign is protracted, the resources of the State will not be equal to the strain.

4. Now, when your weapons are dulled, your ardour damped, your strength exhausted and your treasure spent, other chieftains will spring up to take advantage of your extremity. Then no man, however wise, will be able to avert the consequences that must ensue.

5. Thus, though we have heard of stupid haste in war, cleverness has never been seen associated with long delays.[1]

6. There is no instance of a country having benefited from prolonged warfare.

7. It is only one who is thoroughly acquainted with the evils of war that can thoroughly understand the profitable way of carrying it on.[2]

8. The skilful soldier does not raise a second levy, neither are his supply-waggons loaded more than twice.

9. Bring war material with you from home, but forage on the enemy. Thus the army will have food enough for its needs.

10. Poverty of the State exchequer causes an army to be maintained by contributions from a distance. Contributing to maintain an army at a distance causes the people to be impoverished.

11. On the other hand, the proximity of an army causes prices to go up; and high prices cause the people's substance to be drained away.

12. When their substance is drained away, the peasantry will be afflicted by heavy exactions.

13, 14. With this loss of substance and exhaustion of strength, the homes of the people will be stripped bare, and three-tenths of their income will be dissipated; while Government expenses for broken chariots, worn-out horses, breast-plates and helmets, bows and arrows, spears and shields, protective mantlets, draught-oxen and heavy waggons, will amount to four-tenths of its total revenue.

15. Hence a wise general makes a point of foraging on the enemy. One cartload of the enemy's provisions is equivalent to twenty of one's own, and likewise a single picul of his provender is equivalent to twenty from one's own store.

16. Now in order to kill the enemy, our men must be roused to anger; that there may be advantage from defeating the enemy, they must have their rewards.

17. Therefore in chariot fighting, when ten or more chariots have been taken, those should be rewarded who took the first. Our own flags should be substituted for those of the enemy, and the chariots mingled and used in conjunction with ours. The captured soldiers should be kindly treated and kept.

18. This is called, using the conquered foe to augment one's own strength.

19. In war, then, let your great object be victory, not lengthy campaigns.[3]

20. Thus it may be known that the leader of armies is the arbiter of the people's fate, the man on whom it depends whether the nation shall be in peace or in peril.

III

ATTACK BY STRATAGEM

1. Sun Tzŭ said: In the practical art of war, the best thing of all is to take the enemy's country whole and intact; to shatter and destroy it is not so good. So, too, it is better to recapture an army entire than to destroy it, to capture a regiment, a detachment or a company entire than to destroy them.

2. Hence to fight and conquer in all your battles is not supreme excellence; supreme excellence consists in breaking the enemy's resistance without fighting.

3. Thus the highest form of generalship is to baulk the enemy's plans; the next best is to prevent the junction of the enemy's forces; the next in order is to attack the enemy's army in the field; and the worst policy of all is to besiege walled cities.

4. The rule is, not to besiege walled cities if it can possibly be avoided. The preparation of mantlets,[1] movable shelters, and various implements of war, will take up three whole months; and the piling up of mounds over against the walls will take three months more.

5. The general, unable to control his irritation, will launch his men to the assault like swarming ants, with the result that one-third of his men are slain, while the town still remains untaken. Such are the disastrous effects of a siege.

6. Therefore the skillful leader subdues the enemy's troops without any fighting; he captures their cities without laying siege to them; he overthrows their kingdom without lengthy operations in the field.

7. With his forces intact he will dispute the mastery of the Empire, and thus, without losing a man, his triumph will be complete. This is the method of attacking by stratagem.

8. It is the rule in war, if our forces are ten to the enemy's one, to surround him; if five to one, to attack him; if twice as numerous, to divide our army into two.

9. If equally matched, we can offer battle; if slightly inferior in numbers, we can avoid the enemy; if quite unequal in every way, we can flee from him.

10. Hence, though an obstinate fight may be made by a small force, in the end it must be captured by the larger force.

11. Now the general is the bulwark of the State; if the bulwark is complete at all points, the State will be strong; if the bulwark is defective, the State will be weak.

12. There are three ways in which a ruler can bring misfortune upon his army:—

13. (1) By commanding the army to advance or to retreat, being ignorant of the fact that it cannot obey. This is called hobbling the army.

14. (2) By attempting to govern an army in the same way as he administers a kingdom, being ignorant of the conditions which obtain in an army. This causes restlessness in the soldier's minds.

15. (3) By employing the officers of his army without discrimination, through ignorance of the military principle of adaptation to circumstances. This shakes the confidence of the soldiers.

16. But when the army is restless and distrustful, trouble is sure to come from the other feudal princes. This is simply bringing anarchy into the army, and flinging victory away.

17. Thus we may know that there are five essentials for victory:

(1) He will win who knows when to fight and when not to fight.

(2) He will win who knows how to handle both superior and inferior forces.[2]

(3) He will win whose army is animated by the same spirit throughout all its ranks.

(4) He will win who, prepared himself, waits to take the enemy unprepared.

(5) He will win who has military capacity and is not interfered with by the sovereign.

Victory lies in the knowledge of these five points.

18. Hence the saying: If you know the enemy and know yourself, you need not fear the result of a hundred battles. If you know yourself but not the enemy, for every victory gained you will also suffer a defeat.[3] If you know neither the enemy nor yourself, you will succumb in every battle.

IV

TACTICAL DISPOSITIONS

1. Sun Tzŭ said: The good fighters of old first put themselves beyond the possibility of defeat, and then waited for an opportunity of defeating the enemy.

2. To secure ourselves against defeat lies in our own hands, but the opportunity of defeating the enemy is provided by the enemy himself.

3. Thus the good fighter is able to secure himself against defeat, but cannot make certain of defeating the enemy.

4. Hence the saying: One may *know* how to conquer without being able to *do* it.

5. Security against defeat implies defensive tactics; ability to defeat the enemy means taking the offensive.

6. Standing on the defensive indicates insufficient strength; attacking, a superabundance of strength.

7. The general who is skilled in defence hides in the most secret recesses of the earth; he who is skilled in attack flashes forth from the topmost heights of heaven. Thus on the one hand we have ability to protect ourselves; on the other, a victory that is complete.

8. To see victory only when it is within the ken of the common herd is not the acme of excellence.[1]

9. Neither is it the acme of excellence if you fight and conquer and the whole Empire says, "Well done!"

10. To lift an autumn hair is no sign of great strength; to see the sun and moon is no sign of sharp sight; to hear the noise of thunder is no sign of a quick ear.

11. What the ancients called a clever fighter is one who not only wins, but excels in winning with ease.

12. Hence his victories bring him neither reputation for wisdom nor credit for courage.[2]

13. He wins his battles by making no mistakes. Making no mistakes is what establishes the certainty of victory, for it means conquering an enemy that is already defeated.

14. Hence the skillful fighter puts himself into a position which makes defeat impossible, and does not miss the moment for defeating the enemy.

15. Thus it is that in war the victorious strategist only seeks battle after the victory has been won, whereas he who is destined to defeat first fights and afterwards looks for victory.[3]

16. The consummate leader cultivates the moral law, and strictly adheres to method and discipline; thus it is in his power to control success.

17. In respect of military method, we have, firstly, Measurement; secondly, Estimation of quantity; thirdly, Calculation; fourthly, Balancing of chances; fifthly, Victory.

18. Measurement owes its existence to Earth; Estimation of quantity to Measurement; Calculation to Estimation of quantity; Balancing of chances to Calculation; and Victory to Balancing of chances.

19. A victorious army opposed to a routed one, is as a pound's weight placed in the scale against a single grain.

20. The onrush of a conquering force is like the bursting of pent-up waters into a chasm a thousand fathoms deep.

V

ENERGY

1. Sun Tzŭ said: The control of a large force is the same principle as the control of a few men: it is merely a question of dividing up their numbers.

2. Fighting with a large army under your command is nowise different from fighting with a small one: it is merely a question of instituting signs and signals.

3. To ensure that your whole host may withstand the brunt of the enemy's attack and remain unshaken—this is effected by manœuvres direct and indirect.

4. That the impact of your army may be like a grindstone dashed against an egg—this is effected by the science of weak points and strong.

5. In all fighting, the direct method may be used for joining battle, but indirect methods will be needed in order to secure victory.

6. Indirect tactics, efficiently applied, are inexhaustible as Heaven and Earth, unending as the flow of rivers and streams; like the sun and moon, they end but to begin anew; like the four seasons, they pass away to return once more.

7. There are not more than five musical notes, yet the combinations of these five give rise to more melodies than can ever be heard.

8. There are not more than five primary colours, yet in combination they produce more hues than can ever been seen.

9. There are not more than five cardinal tastes, yet combinations of them yield more flavours than can ever be tasted.

10. In battle, there are not more than two methods of attack—the direct and the indirect; yet these two in combination give rise to an endless series of manœuvres.

11. The direct and the indirect lead on to each other in turn. It is like moving in a circle—you never come to an end. Who can exhaust the possibilities of their combination?

12. The onset of troops is like the rush of a torrent which will even roll stones along in its course.

13. The quality of decision is like the well-timed swoop of a falcon which enables it to strike and destroy its victim.

14. Therefore the good fighter will be terrible in his onset, and prompt in his decision.

15. Energy may be likened to the bending of a crossbow; decision, to the releasing of a trigger.

16. Amid the turmoil and tumult of battle, there may be

seeming disorder and yet no real disorder at all; amid confusion and chaos, your array may be without head or tail, yet it will be proof against defeat.

17. Simulated disorder postulates perfect discipline; simulated fear postulates courage; simulated weakness postulates strength.

18. Hiding order beneath the cloak of disorder is simply a question of subdivision; concealing courage under a show of timidity presupposes a fund of latent energy; masking strength with weakness is to be effected by tactical dispositions.

19. Thus one who is skilful at keeping the enemy on the move maintains deceitful appearances, according to which the enemy will act. He sacrifices something, that the enemy may snatch at it.

20. By holding out baits, he keeps him on the march; then with a body of picked men he lies in wait for him.

21. The clever combatant looks to the effect of combined energy, and does not require too much from individuals. Hence his ability to pick out the right men and utilize combined energy.[1]

22. When he utilises combined energy, his fighting men become as it were like unto rolling logs or stones. For it is the nature of a log or stone to remain motionless on level ground, and to move when on a slope; if four-cornered, to come to a standstill, but if round-shaped, to go rolling down.

23. Thus the energy developed by good fighting men is as the momentum of a round stone rolled down a mountain thousands of feet in height. So much on the subject of energy.

VI

WEAK POINTS AND STRONG

1. Sun Tzŭ said: Whoever is first in the field and awaits the coming of the enemy, will be fresh for the fight; whoever is second in the field and has to hasten to battle will arrive exhausted.

2. Therefore the clever combatant imposes his will on the enemy, but does not allow the enemy's will to be imposed on him.

3. By holding out advantages to him, he can cause the enemy to approach of his own accord; or, by inflicting damage, he can make it impossible for the enemy to draw near.

4. If the enemy is taking his ease, he can harass him; if well supplied with food, he can starve him out; if quietly encamped, he can force him to move.

5. Appear at points which the enemy must hasten to defend; march swiftly to places where you are not expected.

6. An army may march great distances without distress, if it marches through country where the enemy is not.

7. You can be sure of succeeding in your attacks if you only attack places which are undefended. You can ensure the safety of your defence if you only hold positions that cannot be attacked.

8. Hence that general is skilful in attack whose opponent does not know what to defend; and he is skillful in defence whose opponent does not know what to attack.[1]

9. O divine art of subtlety and secrecy! Through you we learn to be invisible, through you inaudible; and hence we can hold the enemy's fate in our hands.

10. You may advance and be absolutely irresistible, if you make for the enemy's weak points; you may retire and be safe from pursuit if your movements are more rapid than those of the enemy.

11. If we wish to fight, the enemy can be forced to an engagement even though he be sheltered behind a high rampart and a deep ditch. All we need do is attack some other place that he will be obliged to relieve.

12. If we do not wish to fight, we can prevent the enemy from engaging us even though the lines of our encampment be merely traced out on the ground. All we need do is to throw something odd and unaccountable in his way.

13. By discovering the enemy's dispositions and remaining invisible ourselves, we can keep our forces concentrated, while the enemy's must be divided.[2]

14. We can form a single united body, while the enemy must split up into fractions. Hence there will be a whole pitted against separate parts of a whole, which means that we shall be many to the enemy's few.

15. And if we are able thus to attack an inferior force with a superior one, our opponents will be in dire straits.

16. The spot where we intend to fight must not be made known; for then the enemy will have to prepare against a possible attack at several different points; and his forces being thus distributed in many directions, the numbers we shall have to face at any given point will be proportionately few.

17. For should the enemy strengthen his van, he will weaken his rear; should he strengthen his rear, he will weaken his van; should he strengthen his left, he will weaken his right; should he strengthen his right, he will weaken his left. If he sends reinforcements everywhere, he will everywhere be weak.

18. Numerical weakness comes from having to prepare against possible attacks; numerical strength, from compelling our adversary to make these preparations against us.

19. Knowing the place and the time of the coming battle, we may concentrate from the greatest distances in order to fight.

20. But if neither time nor place be known, then the left wing will be impotent to succour the right, the right equally impotent to succour the left, the van unable to relieve the rear, or the rear to support the van. How much more so if the furthest portions of the army are anything under a hundred *li* apart, and even the nearest are separated by several *li*![3]

21. Though according to my estimate the soldiers of Yüeh exceed our own in number, that shall advantage them nothing in the matter of victory. I say then that victory can be achieved.

22. Though the enemy be stronger in numbers, we may prevent him from fighting. Scheme so as to discover his plans and the likelihood of their success.

23. Rouse him, and learn the principle of his activity or inactivity. Force him to reveal himself, so as to find out his vulnerable spots.

24. Carefully compare the opposing army with your own, so that you may know where strength is superabundant and where it is deficient.

25. In making tactical dispositions, the highest pitch you can attain is to conceal them; conceal your dispositions, and you will be safe from the prying of the subtlest spies, from the machinations of the wisest brains.

26. How victory may be produced for them out of the enemy's own tactics—that is what the multitude cannot comprehend.

27. All men can see the tactics whereby I conquer, but what none can see is the strategy out of which victory is evolved.

28. Do not repeat the tactics which have gained you one victory, but let your methods be regulated by the infinite variety of circumstances.

29. Military tactics are like unto water; for water in its natural course runs away from high places and hastens downwards.

30. So in war, the way is to avoid what is strong and to strike at what is weak.[4]

31. Water shapes its course according to the nature of the ground over which it flows; the soldier works out his victory in relation to the foe whom he is facing.

32. Therefore, just as water retains no constant shape, so in warfare there are no constant conditions.

33. He who can modify his tactics in relation to his opponent and thereby succeed in winning, may be called a heaven-born captain.

34. The five elements[5] are not always equally predominant; the four seasons make way for each other in turn. There are short days and long; the moon has its periods of waning and waxing.

VII

MANŒUVRING

1. Sun Tzŭ said: In war, the general receives his commands from the sovereign.

2. Having collected an army and concentrated his forces, he must blend and harmonise the different elements thereof before pitching his camp.

3. After that, comes tactical manœuvring, than which there is nothing more difficult. The difficulty of tactical manœuvring consists in turning the devious into the direct, and misfortune into gain.

4. Thus, to take a long and circuitous route, after enticing the enemy out of the way, and though starting after him, to contrive to reach the goal before him, shows knowledge of the artifice of *deviation*.

5. Manœuvring with an army is advantageous; with an undisciplined multitude, most dangerous.

6. If you set a fully equipped army in march in order to snatch an advantage, the chances are that you will be too late. On the other hand, to detach a flying column for the purpose involves the sacrifice of its baggage and stores.

7. Thus, if you order your men to roll up their buff-coats, and make forced marches without halting day or night, covering double the usual distance at a stretch,

doing a hundred *li* in order to wrest an advantage, the leaders of all your three divisions will fall into the hands of the enemy.

8. The stronger men will be in front, the jaded ones will fall behind, and on this plan only one-tenth of your army will reach its destination.

9. If you march fifty *li* in order to outmanœuvre the enemy, you will lose the leader of your first division, and only half your force will reach the goal.

10. If you march thirty *li* with the same object, two-thirds of your army will arrive.

11. We may take it then that an army without its baggage-train is lost; without provisions it is lost; without bases of supply it is lost.

12. We cannot enter into alliances until we are acquainted with the designs of our neighbours.

13. We are not fit to lead an army on the march unless we are familiar with the face of the country—its mountains and forests, its pitfalls and precipices, its marshes and swamps.

14. We shall be unable to turn natural advantage to account unless we make use of local guides.

15. In war, practice dissimulation, and you will succeed. Move only if there is real advantage to be gained.

16. Whether to concentrate or to divide your troops, must be decided by circumstances.

17. Let your rapidity be that of the wind, your compactness that of the forest.

18. In raiding and plundering be like fire, in immovability like a mountain.

19. Let your plans be dark and impenetrable as night, and when you move, fall like a thunderbolt.

20. When you plunder a countryside, let the spoil be divided amongst your men; when you capture new territory, cut it up into allotments for the benefit of the soldiery.

21. Ponder and deliberate before you make a move.

22. He will conquer who has learnt the artifice of deviation. Such is the art of manœuvring.[1]

23. The Book of Army Management[2] says: On the field of battle, the spoken word does not carry far enough: hence the institution of gongs and drums. Nor can ordinary objects be seen clearly enough: hence the institution of banners and flags.

24. Gongs and drums, banners and flags, are means whereby the ears and eyes of the host may be focussed on one particular point.

25. The host thus forming a single united body, is it impossible either for the brave to advance alone, or for the cowardly to retreat alone.[3] This is the art of handling large masses of men.

26. In night-fighting, then, make much use of signal-fires and drums, and in fighting by day, of flags and banners, as a means of influencing the ears and eyes of your army.

27. A whole army may be robbed of its spirit; a commander-in-chief may be robbed of his presence of mind.

28. Now a soldier's spirit is keenest in the morning; by noonday it has begun to flag; and in the evening, his mind is bent only on returning to camp.

29. A clever general, therefore, avoids an army when its spirit is keen, but attacks it when it is sluggish and inclined to return. This is the art of studying moods.

30. Disciplined and calm, to await the appearance of disorder and hubbub amongst the enemy:—this is the art of retaining self-possession.

31. To be near the goal while the enemy is still far from it, to wait at ease while the enemy is toiling and struggling, to be well-fed while the enemy is famished:—this is the art of husbanding one's strength.

32. To refrain from intercepting an enemy whose banners are in perfect order, to refrain from attacking an army drawn up in calm and confident array:—this is the art of studying circumstances.

33. It is a military axiom not to advance uphill against the enemy, nor to oppose him when he comes downhill.

34. Do not pursue an enemy who simulates flight; do not attack soldiers whose temper is keen.

35. Do not swallow bait offered by the enemy. Do not interfere with an army that is returning home.[4]

36. When you surround an army, leave an outlet free. Do not press a desperate foe too hard.

37. Such is the art of warfare.

VIII

VARIATION OF TACTICS

1. Sun Tzŭ said: In war, the general receives his commands from the sovereign, collects his army and concentrates his forces.

2. When in difficult country, do not encamp. In country where high roads intersect, join hands with your allies. Do not linger in dangerously isolated positions. In hemmed-in situations, you must resort to stratagem. In a desperate position, you must fight.

3. There are roads which must not be followed, armies which must be not attacked, towns which must not be besieged,[1] positions which must not be contested, commands of the sovereign which must not be obeyed.

4. The general who thoroughly understands the advantages that accompany variation of tactics knows how to handle his troops.

5. The general who does not understand these, may be well acquainted with the configuration of the country, yet he will not be able to turn his knowledge to practical account.[2]

6. So, the student of war who is unversed in the art of war of varying his plans, even though he be acquainted with the Five Advantages, will fail to make the best use of his men.[3]

7. Hence in the wise leader's plans, considerations of advantage and of disadvantage will be blended together.

8. If our expectation of advantage be tempered in this way, we may succeed in accomplishing the essential part of our schemes.

9. If, on the other hand, in the midst of difficulties we are always ready to seize an advantage, we may extricate ourselves from misfortune.

10. Reduce the hostile chiefs by inflicting damage on them; make trouble for them, and keep them constantly engaged; hold out specious allurements, and make them rush to any given point.

11. The art of war teaches us to rely not on the likelihood of the enemy's not coming, but on our own readiness to receive him; not on the chance of his not attacking, but rather on the fact that we have made our position unassailable.

12. There are five dangerous faults which may affect a general:

(1) Recklessness, which leads to destruction;[4]

(2) cowardice, which leads to capture;[5]

(3) a hasty temper, which can be provoked by insults;[6]

(4) a delicacy of honor which is sensitive to shame;[7]

(5) over-solicitude for his men, which exposes him to worry and trouble.[8]

13. These are the five besetting sins of a general, ruinous to the conduct of war.

14. When an army is overthrown and its leader slain, the cause will surely be found among these five dangerous faults. Let them be a subject of meditation.

IX

THE ARMY ON THE MARCH

1. Sun Tzŭ said: We come now to the question of encamping the army, and observing signs of the enemy. Pass quickly over mountains, and keep in the neighborhood of valleys.[1]

2. Camp in high places, facing the sun. Do not climb heights in order to fight.[2]

3. After crossing a river, you should get far away from it.

4. When an invading force crosses a river in its onward march, do not advance to meet it in mid-stream. It will be best to let half the army get across, and then deliver your attack.[3]

5. If you are anxious to fight, you should not go to meet the invader near a river which he has to cross.

6. Moor your craft higher up than the enemy, and facing the sun. Do not move up-stream to meet the enemy. So much for river warfare.

7. In crossing salt-marshes, your sole concern should be to get over them quickly, without any delay.[4]

8. If forced to fight in a salt-marsh, you should have water and grass near you, and get your back to a clump of trees. So much for operations in salt-marshes.

9. In dry, level country, take up an easily accessible position with rising ground to your right and on your rear, so that the danger may be in front, and safety lie behind. So much for campaigning in flat country.

10. These are the four useful branches of military knowledge[5] which enabled the Yellow Emperor to vanquish four several sovereigns.

11. All armies prefer high ground to low,[6] and sunny places to dark.

12. If you are careful of your men, and camp on hard ground, the army will be free from disease of every kind, and this will spell victory.

13. When you come to a hill or a bank, occupy the sunny side, with the slope on your right rear. Thus you will at once act for the benefit of your soldiers and utilise the natural advantages of the ground.

14. When, in consequence of heavy rains up-country, a river which you wish to ford is swollen and flecked with foam, you must wait until it subsides.

15. Country in which there are precipitous cliffs with torrents running between deep natural hollows, confined places, tangled thickets, quagmires and crevasses, should be left with all possible speed and not approached.

16. While we keep away from such places, we should get the enemy to approach them; while we face them, we should let the enemy have them on his rear.

17. If in the neighbourhood of your camp there should be any hilly country, ponds surrounded by aquatic grass, hollow basins filled with reeds, or woods with thick undergrowth, they must be carefully routed out and searched; for these are places where men in ambush or insidious spies are likely to be lurking.

18 When the enemy is close at hand and remains quiet, he is relying on the natural strength of his position.

19. When he keeps aloof and tries to provoke a battle, he is anxious for the other side to advance.[7]

20. If his place of encampment is easy of access, he is tendering a bait.

21. Movement amongst the trees of a forest shows that the enemy is advancing.[8] The appearance of a number of screens in the midst of thick grass means that the enemy wants to make us suspicious.[9]

22. The rising of birds in their flight is the sign of an ambuscade. Startled beasts indicate that a sudden attack is coming.

23. When there is dust rising in a high column, it is the sign of chariots advancing; when the dust is low, but spread over a wide area, it betokens the approach of infantry. When it branches out in different directions, it shows that parties have been sent to collect firewood. A few clouds of dust moving to and fro signify that the army is encamping.

24. Humble words and increased preparations are signs that the enemy is about to advance. Violent language and driving forward as if to the attack are signs that he will retreat.

25. When the light chariots come out first and take up a position on the wings, it is a sign that the enemy is forming for battle.

26. Peace proposals unaccompanied by a sworn covenant indicate a plot.

27. When there is much running about and the soldiers fall into rank, it means that the critical moment has come.

28. When some are seen advancing and some retreating, it is a lure.

29. When the soldiers stand leaning on their spears, they are faint from want of food.

30. If those who are sent to draw water begin by drinking themselves, the army is suffering from thirst.

31. If the enemy sees an advantage to be gained and makes no effort to secure it, the soldiers are exhausted.

32. If birds gather on any spot, it is unoccupied. Clamour by night betokens nervousness.

33. If there is disturbance in the camp, the general's authority is weak. If the banners and flags are shifted about, sedition is afoot. If the officers are angry, it means that the men are weary.

34. When an army feeds its horses with grain and kills its cattle for food, and when the men do not hang their cooking-pots over the camp-fires, showing that they will not return to their tents, you may know that they are determined to fight to the death.

35. The sight of men whispering together in small knots or speaking in subdued tones points to disaffection amongst the rank and file.

36. Too frequent rewards signify that the enemy is at the end of his resources;[10] too many punishments betray a condition of dire distress.[11]

37. To begin by bluster, but afterwards to take fright at the enemy's numbers, shows a supreme lack of intelligence.

38. When envoys are sent with compliments in their mouths, it is a sign that the enemy wishes for a truce.

39. If the enemy's troops march up angrily and remain facing ours for a long time without either joining battle or taking themselves off again, the situation is one that demands great vigilance and circumspection.

40. If our troops are no more in number than the enemy, that is amply sufficient; it only means that no direct attack can be made. What we can do is simply to concentrate all our available strength, keep a close watch on the enemy, and obtain reinforcements.

41. He who exercises no forethought but makes light of his opponents is sure to be captured by them.

42. If soldiers are punished before they have grown attached to you, they will not prove submissive; and, unless submissive, they will be practically useless. If, when the soldiers have become attached to you, punishments are not enforced, they will still be useless.

43. Therefore soldiers must be treated in the first instance with humanity, but kept under control by means of iron discipline. This is a certain road to victory.

44. If in training soldiers commands are habitually enforced, the army will be well-disciplined; if not, its discipline will be bad.

45. If a general shows confidence in his men but always insists on his orders being obeyed, the gain will be mutual.

X

TERRAIN

1. Sun Tzŭ said: We may distinguish six kinds of terrain, to wit: (1) Accessible ground; (2) entangling ground; (3) temporising ground; (4) narrow passes; (5) precipitous heights; (6) positions at a great distance from the enemy.

2. Ground which can be freely traversed by both sides is called *accessible.*

3. With regard to ground of this nature, be before the enemy in occupying the raised and sunny spots, and carefully guard your line of supplies. Then you will be able to fight with advantage.

4. Ground which can be abandoned but is hard to re-occupy is called *entangling.*

5. From a position of this sort, if the enemy is unprepared, you may sally forth and defeat him. But if the enemy is prepared for your coming, and you fail to defeat him, then, return being impossible, disaster will ensue.

6. When the position is such that neither side will gain by making the first move, it is called *temporising* ground.

7. In a position of this sort, even though the enemy should offer us an attractive bait, it will be advisable

not to stir forth, but rather to retreat, thus enticing the enemy in his turn; then, when part of his army has come out, we may deliver our attack with advantage.

8. With regard to *narrow passes,* if you can occupy them first, let them be strongly garrisoned and await the advent of the enemy.

9. Should the army forestall you in occupying a pass, do not go after him if the pass is fully garrisoned, but only if it is weakly garrisoned.

10. With regard to *precipitous heights,* if you are beforehand with your adversary, you should occupy the raised and sunny spots, and there wait for him to come up.[1]

11. If the enemy has occupied them before you, do not follow him, but retreat and try to entice him away.

12. If you are situated at a great distance from the enemy, and the strength of the two armies is equal, it is not easy to provoke a battle, and fighting will be to your disadvantage.

13. These six are the principles connected with Earth. The general who has attained a responsible post must be careful to study them.

14. Now an army is exposed to six several calamities, not arising from natural causes, but from faults for which the general is responsible. These are: (1) Flight; (2) insubordination; (3) collapse; (4) ruin; (5) disorganisation; (6) rout.

15. Other conditions being equal, if one force is hurled against another ten times its size, the result will be the *flight* of the former.

16. When the common soldiers are too strong and their officers too weak, the result is *insubordination*. When the officers are too strong and the common soldiers too weak, the result is *collapse*.

17. When the higher officers are angry and insubordinate, and on meeting the enemy give battle on their own account from a feeling of resentment, before the commander-in-chief can tell whether or no he is in a position to fight, the result is *ruin*.

18. When the general is weak and without authority; when his orders are not clear and distinct; when there are no fixed duties assigned to officers and men, and the ranks are formed in a slovenly haphazard manner, the result is utter *disorganisation*.

19. When a general, unable to estimate the enemy's strength, allows an inferior force to engage a larger one, or hurls a weak detachment against a powerful one, and neglects to place picked soldiers in the front rank, the result must be *rout*.

20. These are six ways of courting defeat, which must

be carefully noted by the general who has attained a responsible post.

21. The natural formation of the country is the soldier's best ally; but a power of estimating the adversary, of controlling the forces of victory, and of shrewdly calculating difficulties, dangers and distances, constitutes the test of a great general.

22. He who knows these things, and in fighting puts his knowledge into practice, will win his battles. He who knows them not, nor practices them, will surely be defeated.

23. If fighting is sure to result in victory, then you must fight, even though the ruler forbid it; if fighting will not result in victory, then you must not fight even at the ruler's bidding.[2]

24. The general who advances without coveting fame and retreats without fearing disgrace, whose only thought is to protect his country and do good service for his sovereign, is the jewel of the kingdom.

25. Regard your soldiers as your children, and they will follow you into the deepest valleys; look upon them as your own beloved sons, and they will stand by you even unto death.

26. If, however, you are indulgent, but unable to make your authority felt; kind-hearted, but unable to enforce your commands; and incapable, moreover, of quelling disorder: then your soldiers must be likened to spoilt children; they are useless for any practical purpose.[3]

27. If we know that our own men are in a condition to attack, but are unaware that the enemy is not open to attack, we have gone only halfway towards victory.

28. If we know that the enemy is open to attack, but are unaware that our own men are not in a condition to attack, we have gone only halfway towards victory.

29. If we know that the enemy is open to attack, and also know that our men are in a condition to attack, but are unaware that the nature of the ground makes fighting impracticable, we have still gone only halfway towards victory.

30. Hence the experienced soldier, once in motion, is never bewildered; once he has broken camp, he is never at a loss.

31. Hence the saying: If you know the enemy and know yourself, your victory will not stand in doubt; if you know Heaven and know Earth, you may make your victory complete.

XI

THE NINE SITUATIONS

1. Sun Tzŭ said: The art of war recognizes nine varieties of ground: (1) Dispersive ground; (2) facile ground; (3) contentious ground; (4) open ground; (5) ground of intersecting highways; (6) serious ground; (7) difficult ground; (8) hemmed-in ground; (9) desperate ground.

2. When a chieftain is fighting in his own territory, it is dispersive ground.[1]

3. When he has penetrated into hostile territory, but to no great distance, it is facile ground.

4. Ground the possession of which imports great advantage to either side, is contentious ground.

5. Ground on which each side has liberty of movement is open ground.

6. Ground which forms the key to three contiguous states, so that he who occupies it first has most of the Empire at his command, is a ground of intersecting highways.

7. When an army has penetrated into the heart of a hostile country, leaving a number of fortified cities in its rear, it is serious ground.

8. Mountain forests, rugged steeps, marshes and fens—

all country that is hard to traverse: this is difficult ground.

9. Ground which is reached through narrow gorges, and from which we can only retire by tortuous paths, so that a small number of the enemy would suffice to crush a large body of our men: this is hemmed-in ground.

10. Ground on which we can only be saved from destruction by fighting without delay, is desperate ground.[2]

11. On dispersive ground, therefore, fight not. On facile ground, halt not. On contentious ground, attack not.

12. On open ground, do not try to block the enemy's way. On the ground of intersecting highways, join hands with your allies.

13. On serious ground, gather in plunder. In difficult ground, keep steadily on the march.

14. On hemmed-in ground, resort to stratagem. On desperate ground, fight.

15. Those who were called skilful leaders of old knew how to drive a wedge between the enemy's front and rear; to prevent co-operation between his large and small divisions; to hinder the good troops from rescuing the bad, the officers from rallying their men.

16. When the enemy's men were scattered, they prevented them from concentrating; even when their forces were united, they managed to keep them in disorder.

17. When it was to their advantage, they made a forward move; when otherwise, they stopped still.

18. If asked how to cope with a great host of the enemy in orderly array and on the point of marching to the attack, I should say: "Begin by seizing something which your opponent holds dear; then he will be amenable to your will."

19. Rapidity is the essence of war: take advantage of the enemy's unreadiness, make your way by unexpected routes, and attack unguarded spots.

20. The following are the principles to be observed by an invading force: The further you penetrate into a country, the greater will be the solidarity of your troops, and thus the defenders will not prevail against you.

21. Make forays in fertile country in order to supply your army with food.

22. Carefully study the well-being of your men, and do not overtax them. Concentrate your energy and hoard your strength. Keep your army continually on the move, and devise unfathomable plans.

23. Throw your soldiers into positions whence there is no escape, and they will prefer death to flight. If they will face death, there is nothing they may not achieve. Officers and men alike will put forth their uttermost strength.

24. Soldiers when in desperate straits lose the sense of fear. If there is no place of refuge, they will stand firm. If they are in hostile country, they will show a stubborn front. If there is no help for it, they will fight hard.

25. Thus, without waiting to be marshalled, the soldiers will be constantly on the *qui vive*; without waiting to be asked, they will do your will; without restrictions, they will be faithful; without giving orders, they can be trusted.

26. Prohibit the taking of omens, and do away with superstitious doubts. Then, until death itself comes, no calamity need be feared.[3]

27. If our soldiers are not overburdened with money, it is not because they have a distaste for riches; if their lives are not unduly long, it is not because they are disinclined to longevity.[4]

28. On the day they are ordered out to battle, your soldiers may weep, those sitting up bedewing their garments, and those lying down letting the tears run down their cheeks. But let them once be brought to bay, and they will display the courage of a Chu or a Kuei.[5]

29. The skillful tactician may be likened to the *shuai-jan*. Now the *shuai-jan* is a snake that is found in the Ch'ang mountains. Strike at its head, and you will be attacked by its tail; strike at its tail, and you will be attacked by its head; strike at its middle, and you will be attacked by head and tail both.

30. Asked if an army can be made to imitate the *shuai-jan*, I should answer, Yes. For the men of Wu and the men of Yüeh are enemies; yet if they are crossing a river in the same boat and are caught by a storm, they will come to each other's assistance just as the left hand helps the right.[6]

31. Hence it is not enough to put one's trust in the tethering of horses, and the burying of chariot wheels in the ground.

32. The principle on which to manage an army is to set up one standard of courage which all must reach.

33. How to make the best of both strong and weak—that is a question involving the proper use of ground.[7]

34. Thus the skilful general conducts his army just as though he were leading a single man, willy-nilly, by the hand.

35. It is the business of a general to be quiet and thus ensure secrecy; upright and just, and thus maintain order.

36. He must be able to mystify his officers and men by false reports and appearances, and thus keep them in total ignorance.

37. By altering his arrangements and changing his plans, he keeps the enemy without definite knowledge. By shifting his camp and taking circuitous routes, he prevents the enemy from anticipating his purpose.

38. At the critical moment, the leader of an army acts like one who has climbed up a height and then kicks away the ladder behind him. He carries his men deep into hostile territory before he shows his hand.

39. He burns his boats and breaks his cooking-pots; like a shepherd driving a flock of sheep, he drives his men this way and that, and nothing knows whither he is going.

40. To muster his host and bring it into danger:—this may be termed the business of the general.[8]

41. The different measures suited to the nine varieties of ground; the expediency of aggressive or defensive tactics; and the fundamental laws of human nature: these are things that must most certainly be studied.

42. When invading hostile territory, the general principle is, that penetrating deeply brings cohesion; penetrating but a short way means dispersion.

43. When you leave your own country behind, and take your army across neighborhood territory, you find yourself on critical ground. When there are means of communication on all four sides, the ground is one of intersecting highways.

44. When you penetrate deeply into a country, it is serious ground. When you penetrate but a little way, it is facile ground.

45. When you have the enemy's strongholds on your rear, and narrow passes in front, it is hemmed-in ground. When there is no place of refuge at all, it is desperate ground.

46. Therefore, on dispersive ground, I would inspire my men with unity of purpose. On facile ground, I would see that there is close connection between all parts of my army.

47. On contentious ground, I would hurry up my rear.[9]

48. On open ground, I would keep a vigilant eye on my defences. On ground of intersecting highways, I would consolidate my alliances.

49. On serious ground, I would try to ensure a continuous stream of supplies. On difficult ground, I would keep pushing on along the road.[10]

50. On hemmed-in ground, I would block any way of retreat. On desperate ground, I would proclaim to my soldiers the hopelessness of saving their lives.[11]

51. For it is the soldier's disposition to offer an obstinate resistance when surrounded, to fight hard when he cannot help himself, and to obey promptly when he has fallen into danger.

52. We cannot enter into alliance with neighbouring princes until we are acquainted with their designs. We are not fit to lead an army on the march unless we are familiar with the face of the country—its mountains and forests, its pitfalls and precipices, its marshes and swamps. We shall be unable to turn natural advantages to account unless we make use of local guides.

53. To be ignored of any one of the following four or five principles does not befit a warlike prince.

54. When a warlike prince attacks a powerful state, his generalship shows itself in preventing the concentration of the enemy's forces. He overawes his opponents, and their allies are prevented from joining against him.

55. Hence he does not strive to ally himself with all and sundry, nor does he foster the power of other states. He carries out his own secret designs, keeping his antagonists in awe. Thus he is able to capture their cities and overthrow their kingdoms.

56. Bestow rewards without regard to rule, issue orders without regard to previous arrangements; and you will be able to handle a whole army as though you had to do with but a single man.

57. Confront your soldiers with the deed itself; never let them know your design. When the outlook is bright, bring it before their eyes; but tell them nothing when the situation is gloomy.

58. Place your army in deadly peril, and it will survive; plunge it into desperate straits, and it will come off in safety.

59. For it is precisely when a force has fallen into harm's way that is capable of striking a blow for victory.

60. Success in warfare is gained by carefully accommodating ourselves to the enemy's purpose.[12]

61. By persistently hanging on the enemy's flank, we shall succeed in the long run in killing the commander-in-chief.

62. This is called ability to accomplish a thing by sheer cunning.

63. On the day that you take up your command, block the frontier passes, destroy the official tallies, and stop the passage of all emissaries.[13]

64. Be stern in the council-chamber, so that you may control the situation.

65. If the enemy leaves a door open, you must rush in.

66. Forestall your opponent by seizing what he holds dear, and subtly contrive to time his arrival on the ground.[14]

67. Walk in the path defined by rule, and accommodate yourself to the enemy until you can fight a decisive battle.

68. At first, then, exhibit the coyness of a maiden, until the enemy gives you an opening; afterwards emulate the rapidity of a running hare, and it will be too late for the enemy to oppose you.

XII

THE ATTACK BY FIRE

1. Sun Tzŭ said: There are five ways of attacking with fire. The first is to burn soldiers in their camp; the second is to burn stores; the third is to burn baggage-trains; the fourth is to burn arsenals and magazines; the fifth is to hurl dropping fire amongst the enemy.

2. In order to carry out an attack, we must have means available. The material for raising fire should always be kept in readiness.

3. There is a proper season for making attacks with fire, and special days for starting a conflagration.

4. The proper season is when the weather is very dry; the special days are those when the moon is in the constellations of the Sieve, the Wall, the Wing or the Cross-bar;[1] for these four are all days of rising wind.

5. In attacking with fire, one should be prepared to meet five possible developments:

6. (1) When fire breaks out inside to enemy's camp, respond at once with an attack from without.

7. (2) If there is an outbreak of fire, but the enemy's soldiers remain quiet, bide your time and do not attack.

8. (3) When the force of the flames has reached its height, follow it up with an attack, if that is practicable; if not, stay where you are.

9. (4) If it is possible to make an assault with fire from without, do not wait for it to break out within, but deliver your attack at a favourable moment.

10. (5) When you start a fire, be to windward of it. Do not attack from the leeward.

11. A wind that rises in the daytime lasts long, but a night breeze soon falls.

12. In every army, the five developments connected with fire must be known, the movements of the stars calculated, and a watch kept for the proper days.

13. Hence those who use fire as an aid to the attack show intelligence; those who use water as an aid to the attack gain an accession of strength.

14. By means of water, an enemy may be intercepted, but not robbed of all his belongings.

15. Unhappy is the fate of one who tries to win his battles and succeed in his attacks without cultivating the spirit of enterprise; for the result is waste of time and general stagnation.[2]

16. Hence the saying: The enlightened ruler lays his plans well ahead; the good general cultivates his resources.

17. Move not unless you see an advantage; use not your troops unless there is something to be gained; fight not unless the position is critical.

18. No ruler should put troops into the field merely to gratify his own spleen; no general should fight a battle simply out of pique.

19. If it is to your advantage, make a forward move; if not, stay where you are.

20. Anger may in time change to gladness; vexation may be succeeded by content.

21. But a kingdom that has once been destroyed can never come again into being; nor can the dead ever be brought back to life.

22. Hence the enlightened ruler is heedful, and the good general full of caution. This is the way to keep a country at peace and an army intact.

XIII

THE USE OF SPIES

1. Sun Tzŭ said: Raising a host of a hundred thousand men and marching them great distances entails heavy loss on the people and a drain on the resources of the State. The daily expenditure will amount to a thousand ounces of silver. There will be commotion at home and abroad, and men will drop down exhausted on the highways. As many as seven hundred thousand families will be impeded in their labour.

2. Hostile armies may face each other for years, striving for the victory which is decided in a single day. This being so, to remain in ignorance of the enemy's condition simply because one grudges the outlay of a hundred ounces of silver in honours and emoluments, is the height of inhumanity.[1]

3. One who acts thus is no leader of men, no present help to his sovereign, no master of victory.

4. Thus, what enables the wise sovereign and the good general to strike and conquer, and achieve things beyond the reach of ordinary men, is *foreknowledge*.[2]

5. Now this foreknowledge cannot be elicited from spirits; it cannot be obtained inductively from experience, nor by any deductive calculation.

6. Knowledge of the enemy's dispositions can only be obtained from other men.

7. Hence the use of spies, of whom there are five classes: (1) Local spies; (2) inward spies; (3) converted spies; (4) doomed spies; (5) surviving spies.

8. When these five kinds of spy are all at work, none can discover the secret system. This is called "divine manipulation of the threads." It is the sovereign's most precious faculty.

9. Having *local spies* means employing the services of the inhabitants of a district.[3]

10. Having *inward spies,* making use of officials of the enemy.[4]

11. Having *converted spies,* getting hold of the enemy's spies and using them for our own purposes.[5]

12. Having *doomed spies,* doing certain things openly for purposes of deception, and allowing our spies to know of them and report them to the enemy.[6]

13. *Surviving spies,* finally, are those who bring back news from the enemy's camp.[7]

14. Hence it is that which none in the whole army are more intimate relations to be maintained than with spies. None should be more liberally rewarded. In no other business should greater secrecy be preserved.

15. Spies cannot be usefully employed without a certain intuitive sagacity.[8]

16. They cannot be properly managed without benevolence and straightforwardness.

17. Without subtle ingenuity of mind, one cannot make certain of the truth of their reports.[9]

18. Be subtle! be subtle! and use your spies for every kind of business.

19. If a secret piece of news is divulged by a spy before the time is ripe, he must be put to death together with the man to whom the secret was told.

20. Whether the object be to crush an army, to storm a city, or to assassinate an individual, it is always necessary to begin by finding out the names of the attendants, the aides-de-camp, and door-keepers and sentries of the general in command. Our spies must be commissioned to ascertain these.[10]

21. The enemy's spies who have come to spy on us must be sought out, tempted with bribes, led away and comfortably housed. Thus they will become converted spies and available for our service.

22. It is through the information brought by the converted spy that we are able to acquire and employ local and inward spies.

23. It is owing to his information, again, that we can cause the doomed spy to carry false tidings to the enemy.

24. Lastly, it is by his information that the surviving spy can be used on appointed occasions.

25. The end and aim of spying in all its five varieties is knowledge of the enemy; and this knowledge can only be derived, in the first instance, from the converted spy. Hence it is essential that the converted spy be treated with the utmost liberality.

26. Of old, the rise of the Yin dynasty was due to I Chih who had served under the Hsia. Likewise, the rise of the Chou dynasty was due to Lü Ya who had served under the Yin.

27. Hence it is only the enlightened ruler and the wise general who will use the highest intelligence of the army for purposes of spying, and thereby they achieve great results. Spies are a most important element in water, because on them depends an army's ability to move.

ENDNOTES

I. LAYING PLANS

1. Tu Mu alludes to the remarkable story of Ts'ao Ts'ao (A.D. 155–220), who was such a strict disciplinarian that once, in accordance with his own severe regulations against injury to standing crops, he condemned himself to death for having allowed his horse to shy into a field of corn! However, in lieu of losing his head, he was persuaded to satisfy his sense of justice by cutting off his hair. Ts'ao Ts'ao's own comment on the present passage is characteristically curt: "when you lay down a law, see that it is not disobeyed; if it is disobeyed the offender must be put to death."

II. WAGING WAR

1. This concise and difficult sentence is not well explained by any of the commentators. Ts'ao Kung, Li Ch'üan, Mêng Shih, Tu Yu, Tu Mu and Mei Yao-ch'ên have notes to the effect that a general, though naturally stupid, may nevertheless conquer through sheer force of rapidity. Ho Shih says: "Haste may be stupid, but at any rate it saves expenditure of energy and treasure; protracted operations may be very clever, but they bring calamity in their train." Wang Hsi evades the difficulty by remarking: "Lengthy operations mean an army growing old, wealth being expended, an empty exchequer and distress among the people; true cleverness insures against the occurrence of such calamities." Chang Yü says: "So long as victory can be attained, stupid haste is preferable to clever

dilatoriness." Now Sun Tzŭ says nothing whatever, except possibly by implication, about ill-considered haste being better than ingenious but lengthy operations. What he does say is something much more guarded, namely that, while speed may sometimes be injudicious, tardiness can never be anything but foolish—if only because it means impoverishment to the nation.

2. That is, with rapidity. Only one who knows the disastrous effects of a long war can realize the supreme importance of rapidity in bringing it to a close. Only two commentators seem to favor this interpretation, but it fits well into the logic of the context, whereas the rendering, "He who does not know the evils of war cannot appreciate its benefits," is distinctly pointless.

3. As Ho Shih remarks: "War is not a thing to be trifled with." Sun Tzŭ here reiterates the main lesson which this chapter is intended to enforce."

III. ATTACK BY STRATAGEM

1. It is not quite clear what [the word translated as "mantlets", were.] Ts'ao Kung simply defines them as "large shields," but we get a better idea of them from Li Ch'üan, who says they were to protect the heads of those who were assaulting the city walls at close quarters. This seems to suggest a sort of Roman *testudo,* ready made. Tu Mu says they were . . . (wheeled vehicles used in repelling attack, according to K'ang His), but this is denied by Ch'ên Hao. See ch. II, sect. 14. The name is also applied to turrets on city walls. Of the [movable shelters] we get a

fairly clear description from several commentators. They were wooden missile-proof structures on four wheels, propelled from within, covered over with raw hides, and used in sieges to convey parties of men to and from the walls, for the purpose of filling up the encircling moat with earth. Tu Mu adds that they are now called "wooden donkeys."

2. This is not merely the general's ability to estimate numbers correctly, as Li Ch'üan and others make out. Chang Yü expounds the saying more satisfactorily: "By applying the art of war, it is possible with a lesser force to defeat a greater, and *vice versâ*. The secret lies in an eye for locality, and in not letting the right moment slip. Thus Wu Tzŭ says: 'With a superior force, make for easy ground; with an inferior one, make for difficult ground.'"

3. Li Ch'üan cites the case of Fu Chien, prince of Ch'in, who in A.D. 383 marched with a vast army against the Chin Emperor. When warned not to despise an enemy who could command the services of such men as Hsieh An and Huan Ch'ung, he boastfully replied: "I have the population of eight provinces at my back, infantry and horsemen to the number of one million; why, they could dam up the Yangtsze River itself by merely throwing their whips into the stream. What danger have I to fear?" Nevertheless, his forces were soon after disastrously routed at the Fei River, and he was obliged to beat a hasty retreat.

IV. TACTICAL DISPOSITIONS

1. As Ts'ao Kung remarks, "the thing is to see the plant before it has germinated," to foresee the event before the action has begun. Li Ch'üan alludes to the story of Han Hsin who, when about to attack the vastly superior army of Chao, which was strongly entrenched in the city of Ch'êng-an, said to his officers: "Gentlemen, we are going to annihilate the enemy, and shall meet again at dinner." The officers hardly took his words seriously, and gave a very dubious assent. But Han Hsin had already worked out in his mind the details of a clever stratagem, whereby, as he foresaw, he was able to capture the city and inflict a crushing defeat on his adversary."

2. Tu Mu explains this very well: "Inasmuch as his victories are gained over circumstances that have not come to light, the world at large knows nothing of them, and he wins no reputation for wisdom; inasmuch as the hostile state submits before there has been any bloodshed, he receives no credit for courage."

3. Ho Shih thus expounds the paradox: "In warfare, first lay plans which will ensure victory, and then lead your army to battle; if you will not begin with stratagem but rely on brute strength alone, victory will no longer be assured."

V. ENERGY

1. Tu Mu says: "He first of all considers the power of his army in the bulk; afterwards he takes individual talent into account, and uses each men according to his capabilities. He does not demand perfection from the untalented."

VI. WEAK POINTS AND STRONG

1. An aphorism which puts the whole art of war in a nutshell.

2. The conclusion is perhaps not very obvious, but Chang Yü (after Mei Yao-Ch'ên) rightly explains it thus: "If the enemy's dispositions are visible, we can make for him in one body; whereas, our own dispositions being kept secret, the enemy will be obliged to divide his forces in order to guard against attack from every quarter."

3. The Chinese of this last sentence is a little lacking in precision, but the mental picture we are required to draw is probably that of an army advancing towards a given rendezvous in separate columns, each of which has orders to be there on a fixed date. If the general allows the various detachments to proceed at haphazard, without precise instructions as to the time and place of meeting, the enemy will be able to annihilate the army in detail. Chang Yü's note may be worth quoting here: "If we do not know the place where our opponents mean to concentrate or the day on which they will join battle, our unity will be forfeited through our preparations

for defense, and the positions we hold will be insecure. Suddenly happening upon a powerful foe, we shall be brought to battle in a flurried condition, and no mutual support will be possible between wings, vanguard or rear, especially if there is any great distance between the foremost and hindmost divisions of the army."

4. Like water, taking the line of lea st resistance.

5. Water, fire, wood, metal, earth.

VII. MANŒUVRING

1. With these words, the chapter would naturally come to an end. But there now follows a long appendix in the shape of an extract from an earlier book on War, now lost, but apparently extant at the time when Sun Tzŭ wrote. The style of this fragment is not noticeably different from that of Sun Tzŭ himself, but no commentator raises a doubt as to its genuineness.

2. It is perhaps significant that none of the earlier commentators give us any information about this work. Mei Yao-chên calls it "an ancient military classic," and Wang Hsi, "an old book on war." Considering the enormous amount of fighting that had gone on for centuries before Sun Tzŭ's time between the various kingdoms and principalities of China, it is not in itself improbable that a collection of military maxims should have been made and written down at some earlier period.

3. Chang Yü quotes a saying: "Equally guilty are those who advance against orders and those who retreat against orders." Tu Mu tells a story in this connection of Wu Ch'i, when he was fighting against the Ch'in State. Before the battle had begun, one of his soldiers, a man of matchless daring, sallied forth by himself, captured two heads from the enemy, and returned to camp. Wu Ch'i had the man instantly executed, whereupon an officer ventured to remonstrate, saying: "This man was a good soldier, and ought not to have been beheaded." Wu Ch'i replied: "I fully believe he was a good soldier, but I had him beheaded because he acted without orders."

4. The commentators explain this rather singular piece of advice by saying that a man whose heart is set on returning home will fight to the death against any attempt to bar his way, and is therefore too dangerous an opponent to be tackled. Chang Yü quotes the words of Han Hsin: "Invincible is the soldier who hath his desire and returneth homewards." A marvelous tale is told of Ts'ao Ts'ao's courage and resource in the *San Kuo Chi*: In A.D. 198, he was besieging Chang Hsiu in Jang, when Liu Piao sent reinforcements with a view to cutting off Ts'ao's retreat. The latter was obliged to draw off his troops, only to find himself hemmed in between two enemies, who were guarding each outlet of a narrow pass in which he had engaged himself. In this desperate plight Ts'ao waited until nightfall, when he bored a tunnel into the mountain side and laid an ambush in it. As soon as the whole army had passed by, the hidden troops fell on his rear, while Ts'ao himself turned and

met his pursuers in front, so that they were thrown into confusion and annihilated. Ts'ao Ts'ao said afterwards: "The brigands tried to check my army in its retreat and brought me to battle in a desperate position: hence I knew how to overcome them."

VIII. VARIATION OF TACTICS

1. Ts'ao Kung gives an interesting illustration from his own experience. When invading the territory of Hsu-chou, he ignored the city of Hua-pi, which lay directly in his path, and pressed on into the heart of the country. This excellent strategy was rewarded by the subsequent capture of no fewer than fourteen important district cities. Chang Yü says: "No town should be attacked which, if taken, cannot be held, or if left alone, will not cause any trouble." Hsün Ying, when urged to attack Pi-yang, replied: "The city is small and well-fortified; even if I succeed in taking it, it will be no great feat of arms; whereas if I fail, I shall make myself a laughing-stock."

2. Literally, "get the advantage of the ground," which means not only securing good positions, but availing oneself of natural advantages in every possible way. Chang Yü says: "Every kind of ground is characterized by certain natural features, and also gives scope for a certain variability of plan. How it is possible to turn these natural features to account unless topographical knowledge is supplemented by versatility of mind?"

3. Chia Lin . . . tells us that these imply five obvious and generally advantageous lines of action, namely: "if

a certain road is short, it must be followed; if an army is isolated, it must be attacked; if a town is in a parlous condition, it must be besieged; if a position can be stormed, it must be attempted; and if consistent with military operations, the ruler's commands must be obeyed." But there are circumstances which sometimes forbid a general to use these advantages. For instance, "a certain road may be the shortest way for him, but if he knows that it abounds in natural obstacles, or that the enemy has laid an ambush on it, he will not follow that road. A hostile force may be open to attack, but if he knows that it is hard-pressed and likely to fight with desperation, he will refrain from striking," and so on.

4. "Bravery without forethought," as Ts'ao Kung analyzes it, which causes a man to fight blindly and desperately like a mad bull. Such an opponent, says Chang Yü, "must not be encountered with brute force, but may be lured into an ambush and slain."

5. T'ai Kung said: "He who lets an advantage slip will subsequently bring upon himself real disaster." In A.D. 404, Liu Yü pursued the rebel Huan Hsüan up the Yangtsze and fought a naval battle with him at the island of Ch'êng-hung. The loyal troops numbered only a few thousands, while their opponents were in great force. But Huan Hsüan, fearing the fate which was in store for him should be be overcome, had a light boat made fast to the side of his war-junk, so that he might escape, if necessary, at a moment's notice. The natural result was that the fighting spirit of his soldiers

was utterly quenched, and when the loyalists made an attack from windward with fireships, all striving with the utmost ardor to be first in the fray, Huan Hsüan's forces were routed, had to burn all their baggage and fled for two days and nights without stopping.

6. Têng Ch'iang said: "Our adversary is of a choleric temper and easily provoked; let us make constant sallies and break down his walls, then he will grow angry and come out. Once we can bring his force to battle, it is doomed to be our prey." This plan was acted upon, Yao Hsiang came out to fight, was lured as far as San-Yüan by the enemy's pretended flight, and finally attacked and slain.

7. This need not be taken to mean that a sense of honor is really a defect in a general. What Sun Tzŭ condemns is rather an exaggerated sensitiveness to slanderous reports, the thin-skinned man who is stung by opprobrium, however undeserved.

8. Here again, Sun Tzŭ does not mean that the general is to be careless of the welfare of his troops. All he wishes to emphasize is the danger of sacrificing any important military advantage to the immediate comfort of his men. This is a shortsighted policy, because in the long run the troops will suffer more from the defeat, or, at best, the prolongation of the war, which will be the consequence.

IX. THE ARMY ON THE MARCH

1. Chang Yü tells the following anecdote: Wu-tu Ch'iang was a robber captain in the time of the Later Han, and Ma Yüan was sent to exterminate his gang. Ch'iang having found a refuge in the hills, Ma Yüan made no attempt to force a battle, but seized all the favorable positions commanding supplies of water and forage. Ch'iang was soon in such a desperate plight for want of provisions that he was forced to make a total surrender. He did not know the advantage of keeping in the neighborhood of valleys."

2. Not on high hills, but on knolls or hillocks elevated above the surrounding country. So much for mountain warfare.

3. Li Ch'üan alludes to the great victory won by Han Hsin over Lung Chü at the Wei River. In the *Ch'ien Han Shu*, the battle is described as follows: "The two armies were drawn up on opposite sides of the river. In the night, Han Hsin ordered his men to take some ten thousand sacks filled with sand and construct a dam higher up. Then, leading half his army across, he attacked Lung Chü; but after a time, pretending to have failed in his attempt, he hastily withdrew to the other bank. Lung Chü was much elated by this unlooked-for success, and exclaiming: "I felt sure that Han Hsin was really a coward!" he pursued him and began crossing the river in his turn. Han Hsin now sent a party to cut open the sandbags, thus releasing a great volume of water, which swept down and prevented

the greater portion of Lung Chü's army from getting across. He then turned upon the force which had been cut off, and annihilated it, Lung Chü himself being amongst the slain. The rest of the army, on the further bank, also scattered and fled in all directions."

4. Because of the lack of fresh water, the poor quality of the herbage, and last but not least, because they are low, flat, and exposed to attack.

5. Those, namely, concerned with (1) mountains, (2) rivers, (3) marshes, and (4) plains.

6. "High Ground," says Mei Yao-ch'ên, "is not only more agreeable and salubrious, but more convenient from a military point of view; low ground is not only damp and unhealthy, but also disadvantageous for fighting."

7. Probably because we are in a strong position from which he wishes to dislodge us. "If he came close up to us," says Tu Mu, "and tried to force a battle, he would seem to despise us, and there would be less probability of our responding to the challenge."

8. Ts'ao Kung explains this as "felling trees to clear a passage," and Chang Yü says: "Every man sends out scouts to climb high places and observe the enemy. If a scout sees that the trees of a forest are moving and shaking, he may know that they are being cut down to clear a passage for the enemy's march."

9. Tu Yu's explanation, borrowed from Ts'ao Kung's, is as follows: "The presence of a number of screens or sheds in the midst of thick vegetation is a sure sign that the enemy has fled and, fearing pursuit, has constructed these hiding-places in order to make us suspect an ambush." It appears that these "screens" were hastily knotted together out of any long grass which the retreating enemy happened to come across.

10. Because, when an army is hard pressed, as Tu Mu says, there is always a fear of mutiny, and lavish rewards are given to keep the men in good temper.

11. Because in such case discipline becomes relaxed, and unwonted severity is necessary to keep the men to their duty.

X. TERRAIN

1. Chang Yü tells the following anecdote of P'ei Hsing-chien (A.D. 619–682), who was sent on a punitive expedition against the Turkic tribes. "At night he pitched his camp as usual, and it had already been completely fortified by wall and ditch, when suddenly he gave orders that the army should shift its quarters to a hill near by. This was highly displeasing to his officers, who protested loudly against the extra fatigue which it would entail on the men. P'ei Hsing- chien, however, paid no heed to their remonstrances and had the camp moved as quickly as possible. The same night, a terrific storm came on, which flooded their former place of encampment to the depth of over twelve feet. The recalcitrant officers were amazed

at the sight, and owned that they had been in the wrong. 'How did you know what was going to happen?' they asked. P'ei Hsing-chien replied: 'From this time forward be content to obey orders without asking unnecessary questions.' . . . From this it may be seen," Chang Yü continues, "that high and sunny places are advantageous not only for fighting, but also because they are immune from disastrous floods."

2. Huang Shih-kung of the Ch'in dynasty, who is said to have been the patron of Chang Liang and to have written the *San Lueh*, has these words attributed to him: "The responsibility of setting an army in motion must devolve on the general alone; if advance and retreat are controlled from the Palace, brilliant results will hardly be achieved. Hence the god-like ruler and the enlightened monarch are content to play a humble part in furthering their country's cause (*literally*, kneel down to push the chariot wheel)." This means that "in matters lying outside the *zenana,* the decision of the military commander must be absolute." Chang Yü also quote the saying: "Decrees from the Son of Heaven do not penetrate the walls of a camp."

3. Li Ching once said that if you could make your soldiers afraid of you, they would not be afraid of the enemy. Tu Mu recalls an instance of stern military discipline which occurred in A.D. 219, when Lü Mêng was occupying the town of Chiang-ling. He had given stringent orders to his army not to molest the inhabitants nor take anything from them by force. Nevertheless, a certain officer serving under his banner, who happened

to be a fellow-townsman, ventured to appropriate a bamboo hat belonging to one of the people, in order to wear it over his regulation helmet as a protection against the rain. Lü Mêng considered that the fact of his being also a native of Ju-nan should not be allowed to palliate a clear breach of discipline, and accordingly he ordered his summary execution, the tears rolling down his face, however, as he did so. This act of severity filled the army with wholesome awe, and from that time forth even articles dropped in the highway were not picked up.

XI. THE NINE SITUATIONS

1. So called because the soldiers, being near to their homes and anxious to see their wives and children, are likely to seize the opportunity afforded by a battle and scatter in every direction. "In their advance," observes Tu Mu, "they will lack the valor of desperation, and when they retreat, they will find harbors of refuge."

2. Ch'ên Hao says: "to be on 'desperate ground' is like sitting in a leaking boat or crouching in a burning house." Tu Mu quotes from Li Ching a vivid description of the plight of an army thus entrapped: "Suppose an army invading hostile territory without the aid of local guides:—it falls into a fatal snare and is at the enemy's mercy. A ravine on the left, a mountain on the right, a pathway so perilous that the horses have to be roped together and the chariots carried in slings, no passage open in front, retreat cut off behind, no choice but to proceed in single file. Then, before there is time to range

our soldiers in order of battle, the enemy is overwhelming strength suddenly appears on the scene. Advancing, we can nowhere take a breathing-space; retreating, we have no haven of refuge. We seek a pitched battle, but in vain; yet standing on the defensive, none of us has a moment's respite. If we simply maintain our ground, whole days and months will crawl by; the moment we make a move, we have to sustain the enemy's attacks on front and rear. The country is wild, destitute of water and plants; the army is lacking in the necessaries of life, the horses are jaded and the men worn-out, all the resources of strength and skill unavailing, the pass so narrow that a single man defending it can check the onset of ten thousand; all means of offense in the hands of the enemy, all points of vantage already forfeited by ourselves:—in this terrible plight, even though we had the most valiant soldiers and the keenest of weapons, how could they be employed with the slightest effect?"

3. The superstitious, "bound in to saucy doubts and fears," degenerate into cowards and "die many times before their deaths." Tu Mu quotes Huang Shih-kung: "'Spells and incantations should be strictly forbidden, and no officer allowed to inquire by divination into the fortunes of an army, for fear the soldiers' minds should be seriously perturbed.' The meaning is," he continues, "that if all doubts and scruples are discarded, your men will never falter in their resolution until they die."

4. Chang Yü has the best note on this passage: "Wealth and long life are things for which all men have a natural

inclination. Hence, if they burn or fling away valuables, and sacrifice their own lives, it is not that they dislike them, but simply that they have no choice." Sun Tzŭ is slyly insinuating that, as soldiers are but human, it is for the general to see that temptations to shirk fighting and grow rich are not thrown in their way.

5. [Chu] was the personal name of Chuan Chu, a native of the Wu State and contemporary with Sun Tzŭ himself, who was employed by Kung-tzŭ Kuang, better known as Ho Lü Wang, to assassinate his sovereign Wang Liao with a dagger which he secreted in the belly of a fish served up at a banquet. He succeeded in his attempt, but was immediately hacked to pieced by the king's bodyguard. This was in 515 B.C. The other hero referred to, Ts'ao Kuei (or Ts'ao Mo), performed the exploit which has made his name famous 166 years earlier, in 681 B.C. Lu had been thrice defeated by Ch'i, and was just about to conclude a treaty surrendering a large slice of territory, when Ts'ao Kuei suddenly seized Huan Kung, the Duke of Ch'i, as he stood on the altar steps and held a dagger against his chest. None of the duke's retainers dared to move a muscle, and Ts'ao Kuei proceeded to demand full restitution, declaring the Lu was being unjustly treated because she was a smaller and a weaker state. Huan Kung, in peril of his life, was obliged to consent, whereupon Ts'ao Kuei flung away his dagger and quietly resumed his place amid the terrified assemblage without having so much as changed color. As was to be expected, the Duke wanted afterwards to repudiate the bargain, but his wise old counselor Kuan Chung pointed out to him

the impolicy of breaking his word, and the upshot was that this bold stroke regained for Lu the whole of what she had lost in three pitched battles.

6. The meaning is: If two enemies will help each other in a time of common peril, how much more should two parts of the same army, bound together as they are by every tie of interest and fellow-feeling. Yet it is notorious that many a campaign has been ruined through lack of cooperation, especially in the case of allied armies.

7. Mei Yao-Ch'ên's paraphrase is: "The way to eliminate the differences of strong and weak and to make both serviceable is to utilize accidental features of the ground." Less reliable troops, if posted in strong positions, will hold out as long as better troops on more exposed terrain. The advantage of position neutralizes the inferiority in stamina and courage. Col. Henderson, in *The Science of War,* says: "With all respect to the text books, and to the ordinary tactical teaching, I am inclined to think that the study of ground is often overlooked, and that by no means sufficient importance is attached to the selection of positions. . . and to the immense advantages that are to be derived, whether you are defending or attacking, from the proper utilization of natural features."

8. Sun Tzŭ means that after mobilization there should be no delay in aiming a blow at the enemy's heart. . . . Note how he returns again and again to this point. Among the warring states of ancient China, desertion

was no doubt a much more present fear and serious evil than it is in the armies of today.

9. This is Ts'ao Kung's interpretation. Chang Yü adopts it, saying: "We must quickly bring up our rear, so that head and tail may both reach the goal." That is, they must not be allowed to straggle up a long way apart. Mei Yao-Ch'ên offers another equally plausible explanation: "Supposing the enemy has not yet reached the coveted position, and we are behind him, we should advance with all speed in order to dispute its possession."

10. The commentators take this as referring to forage and plunder, not, as one might expect, to an unbroken communication with a home base.

11. Tu Yu says: "Burn your baggage and impedimenta, throw away your stores and provisions, choke up the wells, destroy your cooking-stoves, and make it plain to your men that they cannot survive, but must fight to the death." Mei Yao-Ch'ên says: "The only chance of life lies in giving up all hope of it." This concludes what Sun Tzŭ has to say about "grounds" and the "variations" corresponding to them.

12. Ts'ao Kung says: "Feign stupidity"—by an appearance of yielding and falling in with the enemy's wishes. Chang Yü's note makes the meaning clear: "If the enemy shows an inclination to advance, lure him on to do so; if he is anxious to retreat, delay on purpose that he may carry

out his intention." The object is to make him remiss and contemptuous before we deliver our attack.

13. Either to or from the enemy's country.

14. Ch'ên Hao's explanation . . . is clear enough: "If I manage to seize a favourable position, but the enemy does not appear on the scene, the advantage thus obtained cannot be turned to any practical account. He who intends therefore, to occupy a position of importance to the enemy, must begin by making an artful appointment, so to speak, with his antagonist, and cajole him into going there as well."

XII. THE ATTACK BY FIRE

1. These are, respectively, the 7th, 14th, 27th, and 28th of the Twenty-eight Stellar Mansions, corresponding roughly to Sagittarius, Pegasus, Crater and Corvus.

2. This is one of the most perplexing passages in Sun Tzŭ. . . .Ts'ao Kung says: "Rewards for good service should not be deferred a single day." And Tu Mu: "If you do not take opportunity to advance and reward the deserving, your subordinates will not carry out your commands, and disaster will ensue." . . . For several reasons, however, and in spite of the formidable array of scholars on the other side, I prefer the interpretation suggested by Mei Yao-Ch'ên alone, whose words I will quote: "Those who want to make sure of succeeding in their battles and assaults must seize the favorable moments when they come and not

shrink on occasion from heroic measures: that is to say, they must resort to such means of attack of fire, water and the like. What they must not do, and what will prove fatal, is to sit still and simply hold to the advantages they have got."

XIII. THE USE OF SPIES

1. Sun Tzŭ's arguement is certainly ingenious. He begins by adverting to the frightful misery and vast expenditure of blood and treasure which war always brings in its train. Now, unless you are kept informed of the enemy's condition, and are ready to strike at the right moment, a war may drag on for years. The only way to get this information is to employ spies, and it is impossible to obtain trustworthy spies unless they are properly paid for their services. But it is surely false economy to grudge a comparatively trifling amount for this purpose, when every day that the war lasts eats up an incalculably greater sum. This grievous burden falls on the shoulders of the poor, and hence Sun Tzŭ concludes that to neglect the use of spies is nothing less than a crime against humanity.

2. That is, knowledge of the enemy's dispositions, and what he means to do.

3. Tu Mu says: "In the enemy's country, win people over by kind treatment, and use them as spies."

4. Tu Mu enumerates the following classes as likely to do good service in this respect: "Worthy men who

have been degraded from office, criminals who have undergone punishment; also, favourite concubines who are greedy for gold, men who are aggrieved at being in subordinate positions, or who have been passed over in the distribution of posts, others who are anxious that their side should be defeated in order that they may have a chance of displaying their ability and talents, fickle turncoats who always want to have a foot in each boat. Officials of these several kinds," he continues, "should be secretly approached and bound to one's interests by means of rich presents. In this way you will be able to find out the state of affairs in the enemy's country, ascertain the plans that are being formed against you, and moreover disturb the harmony and create a breach between the sovereign and his ministers."

5. By means of heavy bribes and liberal promises detaching them from the enemy's service, and inducing them to carry back false information as well as to spy in turn on their own countrymen.

6. Tu Yu gives the best exposition of the meaning: "We ostentatiously do things calculated to deceive our own spies, who must be led to believe that they have been unwittingly disclosed. Then, when these spies are captured in the enemy's lines, they will make an entirely false report, and the enemy will take measures accordingly, only to find that we do something quite different. The spies will thereupon be put to death."

7. Tu Mu says: "Your surviving spy must be a man of keen intellect, though in outward appearance a fool; of shabby exterior, but with a will of iron. He must be active, robust, endowed with physical strength and courage; thoroughly accustomed to all sorts of dirty work, able to endure hunger and cold, and to put up with shame and ignominy."

8. Mei Yao-Ch'ên says: "In order to use them, one must know fact from falsehood, and be able to discriminate between honesty and double-dealing."

9. Mei Yao-Ch'ên says: "Be on your guard against the possibility of spies going over to the service of the enemy."

10. As the first step, no doubt towards finding out if any of these important functionaries can be won over by bribery.